THE EVOLUTION OF REPRODUCTIVE STRATEGIES

THE EVOLUTION OF REPRODUCTIVE STRATEGIES

PROCEEDINGS OF
A ROYAL SOCIETY DISCUSSION MEETING
HELD ON 17 AND 18 OCTOBER 1990

ORGANIZED AND EDITED BY
P. H. HARVEY, L. PARTRIDGE AND
T. R. E. SOUTHWOOD

LONDON
THE ROYAL SOCIETY
1991

Printed in Great Britain for the Royal Society
by the
University Press, Cambridge

First published in *Philosophical Transactions of the Royal Society of London.*
series B, volume 332 (no. 1262), pages 1–104

The text paper used in this publication is alkaline sized with a coating which is predominantly calcium carbonate. The resultant surface pH is in excess of 7.5, which gives maximum practical permanence.

British Library Cataloguing in Publication Data

The evolution of reproductive strategies: a discussion
organized and edited by P. H. Harvey, L. Partridge and T. R. E. Southwood
1. Reproduction. Evolution
I. Harvey, Paul H. II. Partridge, L. III. Southwood, Sir Richard
IV. Royal Society
574.16
ISBN 0-85403-432-3

Published by the Royal Society
6 Carlton House Terrace, London SW1Y 5AG

PREFACE

Many observers, from Aristotle and probably before him, have pondered the profligacy of Nature as exemplified by the great fecundity of the herring or the blowfly, compared with its careful, almost miserly approach evidenced in the care that the stickleback or eagle show for their young. This variation in fecundity is but one facet of the diversity of life-history strategies. Reproduction is a particularly key parameter – it is, after all, the counterpart of the summation of all the losses that occur at every other moment in the life history. So looking at reproductive strategies gives us all the information on one side of the coin and on the size of the coin itself.

An evolutionary perspective on life-history strategies provides an interesting insight. The principle of natural selection means, to a good approximation, that organisms should have evolved to leave as many reproductive offspring as possible. Why then is our world not carpeted with Darwinian Demons that start producing copious offspring at birth, and live forever? The answer must be that reproduction costs, either in terms of future reproductive success or survival. There are a number of key trade-offs, such as late survival versus large numbers of offspring, migration versus early reproduction, and maternal care versus high mortality. In these few examples we see the interacting contributions of numbers, time and space to understanding life-history strategies. The papers by Partridge & Sibly and by Kirkwood & Rose in this volume survey our understanding of such trade-offs, which lie at the heart of contemporary evolutionary theory.

The life histories that we observe today have evolved in response to the selective forces, generated by the habitat in which the organism has lived; Mueller's contribution provides perhaps the most comprehensive experimental analysis of one suggested selective spectrum, *r* versus *K* selection. At the other analytical extreme Harvey & Keymer's survey shows how life-history diversity can be studied using Darwin's favoured technique, the comparative method. The method has changed almost beyond recognition since Darwin's day but the questions that can be tackled have not: cross-taxonomic relations among components of life histories, such as clutch size and longevity, and lifestyle or habitat, help us to understand the adaptive foundation of life-history differences. Charnov & Berrigan complement this wide-ranging biological perspective by producing a comprehensive analytical approach, incorporating all the different components of fecundity and mortality in dimensional analyses that use numbers, time and mass. Such analyses show that the optimal strategy is not just a matter of a single optimal combinations of numbers, time and mass; every species is a prisoner of its evolutionary history. As Buss & Blackstone's experiments may eventually reveal, there exist certain constraints on what can or cannot evolve for any particular phenotype or genotype. Indeed, as Gross's elegant field studies show, it is even possible for different individuals in a population to exhibit evolutionarily stable alternative reproductive strategies, which are dictated by an organism's biology.

Different strategies within a population are also beautifully analysed in Godfray & Parker's study of parent–offspring conflict. As many mothers learn to their emotional cost, offspring frequently prefer to be suckled to a later age than mother's are willing to entertain. It may be little recompense that evolutionary analysts predict exactly that conflict between mothers and their young. Just as the offspring may one day itself become a parent, so many organisms change their reproductive strategies as life proceeds, and often as the environment changes. Bell & Koufopanou's wide-ranging studies of small eukaryotes are beginning to reveal when and why multicellularity and sexuality are evolutionarily favoured within a species; such intraspecific studies control for many variables that are likely to confound the results of cross-species analyses. The allocation of resources to sexuality, and to different sex functions is also the tropic of recent investigation among flowering plants; Charlesworth & Morgan's timely review furthers our understanding of the selective pressures that result in different resource allocation decisions.

As the articles in this volume show, there are patterns to be recognized in life-history strategies, patterns based on the habitat and evolutionary history of organisms. Recognizing and understanding such patterns brings order to the near-infinite richness and complexity of the living world, just as the Periodic Table brought order, the possibility of prediction, and the indications of the underlying mechanisms of the chemical world. This is surely a challenge for our subject. If we are to play our part in practical matters as well as in theoretical science, we cannot allow ourselves to simply bask with pleasure reflecting on the seemingly endless variety of nature, and on those peculiar cases that so intrigue us as naturalists. The papers included in this volume show the progress that is being and has been made to elucidate the patterns, the trade-offs, and the underlying mechanisms of life-history variation in the natural world.

February 1991

Richard Southwood
Paul Harvey
Linda Partridge

CONTENTS

Constraints in the evolution of life histories

LINDA PARTRIDGE[1] AND RICHARD SIBLY[2]

[1] *Institute of Cell, Animal and Population Biology, University of Edinburgh, Zoology Building, West Mains Road, Edinburgh EH9 3JT, U.K.*
[2] *Department of Pure and Applied Zoology, University of Reading, PO Box 228, Reading RG6 2AJ, U.K.*

SUMMARY

The life history favoured by natural selection maximizes fitness, and this implies maximization of fecundity and survival at all ages. The observed diversity in life histories suggests that there are constraints on what can be achieved in practice. Functional constraints occur if only certain combinations of age-specific fertility and survival are possible, either because of the physiology of the organism or because of the ecological impact of its environment. The resulting constrained optimization means that the organism is involved in making trade-offs between life-history characters. A major task for the future is the measurement of trade-off functions in the environment in which the life-history evolved. Natural variation between individuals and populations, genetic studies and experimental manipulations have all been used to detect trade-offs. The last two methods are the most satisfactory, and can be complementary. Experimental manipulations are at their best when based on sound physiological understanding of the traits under manipulation. Constraints can also operate on the long-term. Local optima, evolutionary lags and irreversible evolution may all have contributed to the diversity of life histories.

1. INTRODUCTION

The life history of an organism is the combination of age-specific survival probabilities and fecundities it displays in its natural environment. The physical forms of life cycles are extraordinarily variable, and different creatures march through their principal life events to very different drum-beats. Lifespan and fecundity vary over several orders of magnitude, and the age of first reproduction, itself highly variable, can mark the beginning of a period of repeated breeding that occupies most of the lifespan, or may instead herald a single, suicidal burst (Cole 1954; Stearns 1976).

It is the job of evolutionary biologists to make sense of this kind of diversity. Because the extent and timing of progeny production are characters very close to fitness itself (see, for example, Gustaffson 1986), life histories are of special interest, and raise some of the most challenging issues in evolutionary biology (Caswell 1989); what exactly is fitness, and what are the constraints on the characters and combinations of characters contributing to it? The connection between the life history and fitness is much closer than for characters more removed from demography, and any suspicion that life-history evolution may simply be an epiphenomenon of genetic drift or of selection on other characters can therefore be dismissed. There is therefore a real opportunity both to formulate explicit models of selection and to test their ability to explain the diversity of life histories we see in nature.

Many models of life-history evolution and tests of them have used optimality theory (Parker & Maynard Smith 1990). This approach to the study of adaptation specifies the options open to an organism, defines an optimization criterion as close as possible to Darwinian fitness, assigns pay-offs to the different options and deduces the optimal solution. This procedure has been accused both of circularity, if models are altered in the light of a mismatch with the data, and of ignoring constraints on the evolution of perfection by natural selection (Gould & Lewontin 1979). However, any model must be altered if it fails to fit the data. Furthermore, in many optimality models the idea of constraint is inherent in the way that the options are specified. For instance, optimality theories of life-history evolution explicitly include the notion of functional constraints (Charlesworth 1990), which means that only certain constrained combinations of the characters individually increasing fitness can be realized in practice.

An important area of debate for life historians has been the importance of genetics. There is a built-in assumption in optimality theory that there can exist genotypes capable of producing the optimal phenotype, and that such genotypes will evolve. On the other hand, some geneticists have voiced the suspicion that the inclusion of genetic details in the theoretical models may alter the outcome (see, for example, Charlesworth 1980; Lande 1982; Rose *et al.* 1987). Recent theoretical work using quantitative genetics (Charnov 1989; Charlesworth 1990) has suggested that optimality and genetic approaches produce similar equilibrium solutions, even with realistic assumptions about the effects of mutation (Charlesworth 1990). Furthermore, many life-history characters do have a demonstrated genetic basis (see Mueller & Ayala 1981; Rose 1984; Gustaff-

Phil. Trans. R. Soc. Lond. B (1991) **332**, 3–13
Printed in Great Britain

son 1986; Caswell 1989). There has also been a related debate about the need to use genetic techniques in empirical studies of life histories. This topic is considered in some detail later in the present paper.

Optimality models and tests have indeed proved powerful in several cases. For instance, almost all of our understanding of the evolution of avian clutch size (which has been shown to be heritable in some species (see van Noordwijk *et al.* 1980; Gustaffson 1986)) is based on optimality reasoning (see, for example, Charnov & Krebs 1974; Gustaffson & Sutherland 1988; van Noordwijk & de Jong 1986; Pettifor *et al.* 1988). However, it is clear that in some cases a genetic approach to theory and testing is essential. To understand the evolution of date of egg-laying in many birds, it is necessary to make a sharp distinction between genetic and purely environmental variation in the character, and to understand the origin of each (Price *et al.* 1988). Furthermore, gene flow between different populations can prevent optimization of clutch size within each (Dhondt *et al.* 1991).

Functional constraints on evolution are not the only ones. Organisms can become stuck in absorbing states; evolution is not necessarily reversible when selective forces return to an earlier form (Bull & Charnov 1985). Evolutionary history can also be important and there are constraints from design. They will all be discussed below.

2. NATURAL SELECTION ON LIFE HISTORIES

The consequences of natural selection acting on life-history variation within populations are well understood. If the age-specific fecundities and survival probabilities of all the individuals in a population are identical, and if no environmental change intervenes to alter these vital rates, then the population will come to have a stable age-structure, and the numbers of individuals present will change at a rate r, called the intrinsic rate of increase or Malthusian parameter (Cole 1954). There is general agreement that natural selection will favour the genotype that maximizes the value of r under the prevailing ecological conditions. r is determined by the life history, a generalized version of which is illustrated in figure 1. For a genotype producing this life history, r is defined by the Euler–Lotka equation:

$$1 = \tfrac{1}{2}\sum_{x} l_x m_x e^{-rt_x}, \qquad (1)$$

where:

t_x = age at the xth breeding attempt

l_x = survival probability to age t_x

m_x = fecundity at age t_x

(Schaffer 1974; Charlesworth & Leon 1976; Charlesworth 1980; Lande 1982; Caswell 1989; see Sibly (1989) for an exposition that defines r for a gene, and shows that selection favours the gene that maximizes the value of r under the prevailing ecological conditions). r is a measure of the fitness of the genotype in the environment in which its life history was measured.

Assuming survival is not associated with age at breeding, timing of offspring production influences r only if population numbers are changing; if they are increasing, offspring produced early in the lifespan are more valuable than those produced late, while if population numbers are decreasing the opposite is true (Cole 1954). However, with the more realistic assumption that there is mortality throughout the juvenile and adult periods, it is always advantageous to produce offspring as early as possible, unless this induces a change in mortality rate (Sibly & Calow 1987). If population numbers are stable, average lifetime reproductive success is a sufficient measure of the fitness of a genotype (Lande 1982; Sibly 1989; Charnov, this symposium). There is justification for the use of this measure because in practice, for a species to persist over evolutionary time, its numbers cannot decline indefinitely, and none are in a permanent state of increase. Some life-history variables may none the less be under selection only during phases of population increase or decline. For instance, in many multivoltine temperate insects, the summer is a period of increase in numbers, when early offspring will be more valuable, while the decrease in numbers from late summer onwards may occur through adult mortality unrelated to life-history characteristics. The Euler–Lotka equation for r automatically includes the effects of population increase or decrease, because r is defined under specified environmental conditions.

Certain simplifying assumptions are necessary for either r or lifetime reproductive success to be accepted as measures of fitness. Perhaps the most important is that fitnesses are not frequency dependent. This assumption is clearly not true where the presence or frequency of one life-history morph affects the way that selection acts on another (see, for example, Gross 1985, this symposium), and then an ESS approach is necessary (Maynard Smith 1982; Parker & Maynard Smith 1990). The theory also assumes that individuals of the same genotype are identical and invariably in the same ecological environment, aspects of which depend on conspecifics; to the extent that individuals, environments and population parameters vary, the theory needs modification to include the effects of temporally- and spatially-variable selection (Orzack & Tuljapurkar 1989). Other assumptions are that mating is random with respect to life-history characters and to age, and that parent–offspring conflict is not limiting parental options (Godfray & Parker, this volume).

The life-history characters maximizing r were described by Cole (1954), and by Law (1979) who dubbed the hypothetical creature displaying them a 'Darwinian Demon'; it commenced reproduction at birth, and produced copious offspring during frequent breeding attempts throughout its infinite lifespan. If such a life history could be realized in practice then it would be universal. The observed diversity in life histories suggests that natural selection cannot be the only contributing factor, and that other forces, namely constraints, must be at work.

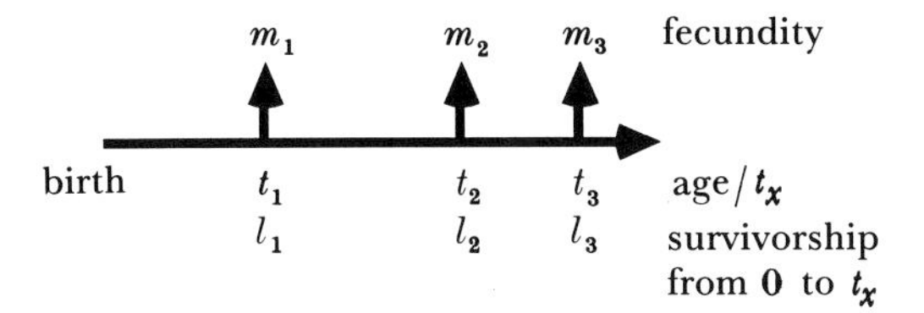

Figure 1. A general life history. The horizontal axis shows the individual's age.

3. FUNCTIONAL CONSTRAINTS

(*a*) *Ecological and physiological constraints*

In the study of life histories, most attention has been directed to the idea that only constrained combinations of the individual life-history characters contributing to r are possible, so that they cannot all be simultaneously maximized by natural selection. There seem to be two main reasons (Calow 1979). First, all organisms live in environments that are to some extent hazardous because of physical risks and the presence of biological enemies. Characters that would increase r can also increase the impact of these hazards, leading to ecological costs. For instance, male frogs *Physalaemus pustulosus* call to attract females, but are more vulnerable to predation by fringe-lipped bats *Trachops cirrhosus* when calling than when silent (Tuttle & Ryan 1981), leading to a survival cost of reproduction. Secondly, physiological costs can also occur if functional constraints are produced by processes internal to the organism. For instance, reproduction may lead to the diversion of resources away from repair or storage, causing a decline in survival or future fecundity (Fisher 1958; Williams 1966). This kind of reproductive cost has been particularly well studied in birds (Partridge 1989*a*; Linden & Moller 1989; Lessells 1991). An example is the demonstration that female collared flycatchers *Ficedula albicollis* allocated enlarged broods in their first year subsequently showed permanently lowered fecundity compared with control females (Gustaffson & Part 1990). Costs of reproduction can therefore affect both sexes, with an effect on mortality or fecundity, and can be exerted instantaneously (in the case of mortality) or with a time delay or both.

Functional constraints can also occur between the juvenile and adult period, although this type of effect is less well explored. For instance, the lifespan and late-life fecundity of female fruitflies *Drosophila melanogaster* can be increased by repeatedly breeding from adults late in their lives. However, this form of artificial selection can also result in a lengthening of the juvenile period and a reduction in larval survival (L. Partridge & K. Fowler, unpublished data). Physiological constraints on parents can also affect offspring characters; it may be impossible for a parent both to produce many offspring and to devote high levels of time or resources to each (Godfray & Parker, this symposium; Harvey & Keymer, this symposium).

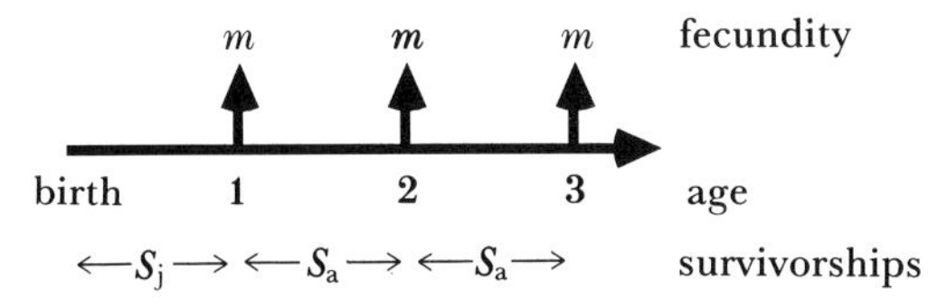

Figure 2. A simple life history first analysed by Schaffer (1974) (cf. figure 1). For this life history, equation (1) simplifies to $e^r = S_a + \frac{1}{2} S_j m$, where S_a and S_j are the adult and juvenile survivorship, respectively.

Organisms are therefore forced to reach some sort of compromise between the demands of competing costly activities, and the nature of that compromise is known as a trade-off. The nature of the constraints will vary between different taxa. These functional constraints are explicitly included in the theory of life-history evolution.

(*b*) *Functional constraints in the theory*

Williams (1966) used Fisher's (1958) idea of reproductive value to explore theoretically the consequences of a conflict between current and future reproduction, caused either by an increased likelihood of death of the parent, or by a negative effect on future fecundity. His important conclusion was that, depending upon the exact relation between possible combinations of current and future reproduction, it could pay to reduce current fecundity, because the loss of progeny would be more than compensated by the improvement in the prospects for reproduction in the future. The critical determinants of the optimal decision are the relative value of adults and progeny, and the function linking the possible combinations of current and future reproduction (Gadgil & Bossert 1970; Schaffer 1974; Pianka & Parker 1975; Sibly & Calow 1986).

These can be illustrated using the modified life history shown in figure 2. For any pair (or more) of life-history characters, the options open to an individual can be described using what have been variously known as strategy sets (Maynard Smith 1978; Parker & Maynard Smith 1990), options sets (Sibly 1991), in a modified form as fitness sets (Levins 1968; Pianka & Parker 1975), and which can sometimes be derived from the principle of allocation (Cody 1966). All refer to an n-dimensional space defined by a set of orthogonal axes, each representing one life-history character. Only a restricted region of this space contains character combinations which the organism is capable of achieving in its natural environment. Two possible functions are illustrated in figure 3*a* and *b* for fecundity at age x and adult survivorship in the life history shown in figure 2. The best combination or trade-off for the organism to adopt can be determined by the use of fitness contours along which r is constant and which are plotted in the same space as the options set. In figure 3 these are straight lines of negative slope S_j, the juvenile survival rate, with r increasing as lines are crossed moving away from the origin. In figure 3*a* the options set is drawn as convex out, as might occur if each increment of reproduction eats increasingly further into a budget for personal maintenance. The trade-off leading to the highest value of r (*) is given by the point of contact of the options set with the outermost fitness contour it touches, leading in this case to intermediate values of reproduction and survival

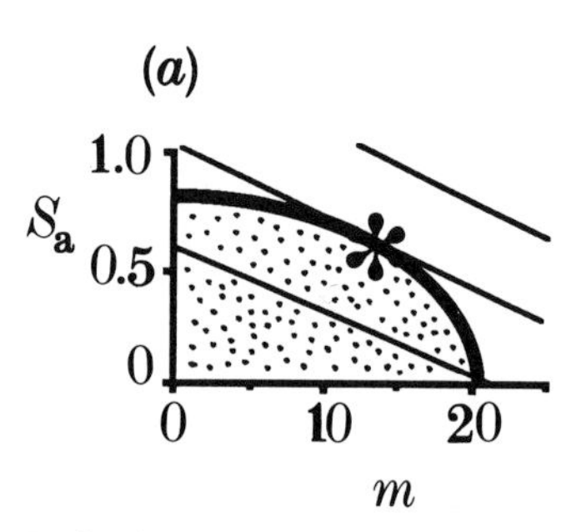

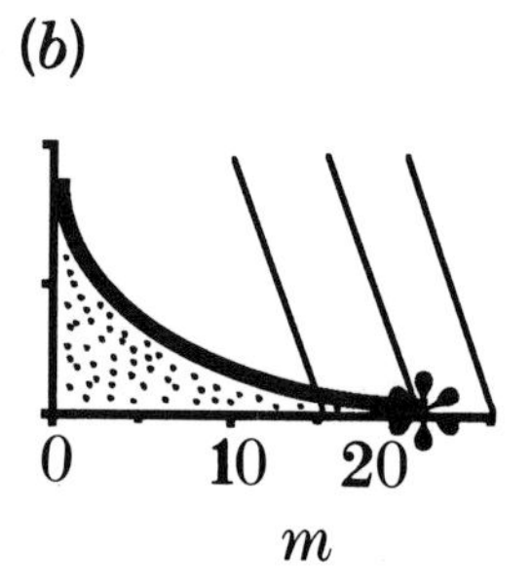

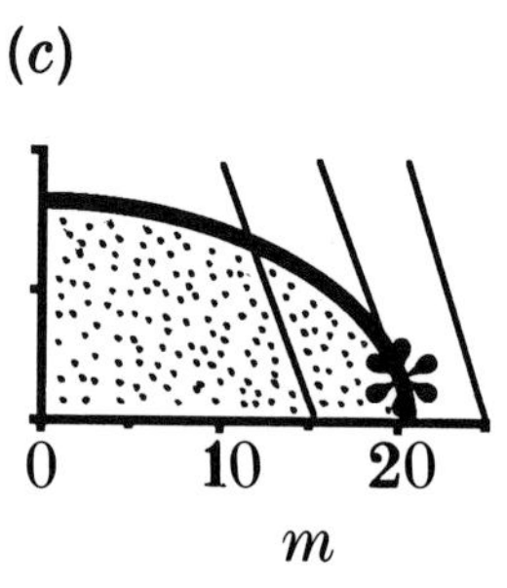

Figure 3. Options sets (dotted) with associated trade-off curves, which are their boundaries, for the life-history variables m and S_a defined in figure 2. In (*a*) and (*c*) the options sets are convex out, in (*b*) it is concave out (see text for biological interpretation). Fitness contours have been superimposed (straight lines), calculated from the equation $S_a = e^r - \frac{1}{2}S_j m$ (see figure 2) and so have slope $-\frac{1}{2}S_j$. * denotes optimal trade-off. Note that low juvenile survivorship selects for iteroparity (*a*), whereas high juvenile survivorship selects for semelparity (*c*).

(iteroparity). If the options set is concave out (figure 3*b*), as might occur if the marginal cost of each offspring reduces with increasing fecundity, suicidal levels of reproduction are optimal (semelparity). The effect of increasing the level of juvenile survival when the options set is convex out (figure 3*c*) is to increase the optimal reproductive rate despite the resulting lower adult survival. This theoretical prediction has recently been elegantly confirmed in field introduction experiments with guppies *Poecilia reticulata* (Reznick *et al.* 1990). If the trade-off varies in curvature over its range, multiple optima may be possible, which could itself explain some life-history diversity.

An important conclusion to emerge from optimality theory is that the shape of the options set is critical in determining the form of the life history. This conclusion is not apparent from quantitative genetic theory, in which the options set is represented by the genetic variance-covariance or G matrix. Genetic models assume linearity of the G matrix in the region of evolutionary equilibrium. In this way the quantitative genetic theories obscure an object of interest, namely the shapes of the trade-off curves (Charnov 1989).

The theory of life-history evolution is well developed. The current state of empirical work is less advanced, and the theory has outstripped our ability to test it. A major difficulty is that we know almost nothing about the shapes of the options sets that determine the possible shapes of trade-off curves.

4. MEASURING OPTIONS SETS

To understand the evolution of life histories we need to know, for specific organisms in specified environments, the set of achievable combinations of different life-history variables. A start in this direction has been made by the detection of the presence of ecological and physiological constraints between pairs of variables; some examples have already been mentioned. Ultimately we wish to extend this knowledge to cases where all the life-history variables and the shapes of the surfaces relating their best achievable values are understood, in at least a few organisms. We need to know where the most important restrictions occur, why they occur and what they look like. The options sets can then be related to the fitness contours for the life histories being studied, to determine if they explain the different trade-off values seen.

There has been much discussion about the appropriate methods for measuring options sets. Is it sufficient to make observations on undisturbed populations? Must we work in the field? What is the best method for an experimental approach? Is it always necessary to do genetic studies? A point of general importance is that the circumstances under which the options set is measured should be those in which the life history evolved, including the impact of ecological hazards and the population density of the study species. These criteria are hard to satisfy, and raise the spectre that we may be dealing with a theory couched in terms of unmeasurable relations. Only time will tell, but a simple conclusion to energe from the approaches about to be discussed is that we need to understand much more about mechanisms if we are to make headway with measuring options sets.

(*a*) *Theoretical considerations*

Options sets can be defined theoretically for some optimization problems (Parker & Maynard Smith 1990). For instance, where male and female progeny are of equal cost and value to their parents, the options set for the proportion of males versus proportion of females in the progeny would be a straight line of negative slope 1. For sex determination systems such as haplo-diploidy, any of the points along the line could also be achieved in practice. Unfortunately, most life-history options sets cannot be so easily defined.

Tantalizingly, it might be possible to identify a *physiological* options set in this way using Cody's (1966) Principle of Allocation. An organism's life history can be viewed at the output of three biological processes, namely maintenance (which affects survival), growth (which affects the timing of reproduction) and reproduction, and these three processes are in competition for nutrients. The Principle of Allocation asserts that, in accordance with the conservation of matter, nutrients allocated to one process are not available to the others. It follows that the options regarding allocations between the three biological processes lie in a plane in the three-dimensional space, the axes of which represent the allocations to each process. So far so good, but how do we get from this set to the life-history variables m_x, l_x and t_x? The relations with m_x and t_x are relatively straightforward (Calow & Sibly 1990), and indeed there have been many measurements of 'reproductive effort', defined as the fraction of the total amount of resources of time and energy available to the individuals at a given age that is devoted to reproduction (Tinkle 1969; Gadgil & Bossert 1970; Hirshfield & Tinkle 1975). The problems arise with the relation between maintenance and mortality rate. Little can be said in general except that the relation is likely to be decreasing and nonlinear. Thus, reducing allocation to maintenance may increase mortality rate gradually until a threshold at which any further increase would pose a serious risk to life. In short, the gain and cost functions for the life-history variables affected by altered nutrient allocation may often be nonlinear. A similar problem occurs when the trade-off spans two generations. For instance, in birds that feed their dependent young, an increase in clutch size is eventually likely to be accompanied by a law of diminishing returns in the number of young fledged, because the total amount of food that can be delivered to the brood is limited by the parents' ability to forage (Lack 1954; Charnov & Krebs 1974). Another problem with this approach is that only some nutrients may be limiting, and these only at particular times. Lastly, the approach concerns itself only with different ways of dividing the cake of available resources, but there is no reason why the size of the cake should not be increased, albeit at some ecological cost to the individual. Such a cost, however, necessarily brings into play new trade-offs, the impact of which cannot be deduced from the Principle of Allocation alone.

An empirical approach to options sets is, therefore, usually desirable. The key question is: what kind of empirical approach? The aim is to persuade an organism to operate over the possible range of values for one life-history variable, and to give its corresponding best possible performance on another, using all the developmental, physiological and behavioural machinery at its disposal. The empirical studies so far described all involved testing the null hypothesis of absence of functional constraints, and all rejected it. Indeed, trade-offs are a logical necessity in a world of risk and finite resources (Stearns 1989). The important questions are whether it is possible empirically to deal directly with the constraints imposed by ecological hazards and by finite resources, and so to measure the shapes of options sets. Possible techniques for doing so will now be considered.

(*b*) *Individual variation*

A method that has frequently been used to detect costs of reproduction is to record data on the naturally occurring variation between the individuals within a population under field or laboratory conditions. The data can be used to test, for instance, for an association across individuals between reproduction and subsequent survival or fecundity. Examples of the use of this method include a demonstration that red deer *Cervus elaphus* females giving birth to calves in one season had lower subsequent survival and fecundity than non-reproductive females (Clutton-Brock *et al.* 1982); work showing that subsequent fecundity is higher in female collared flycatchers *F. albicollis* breeding for their first time at the age of 2 years than in females breeding in their first year (Gustaffson & Part 1990); and the study already mentioned (Tuttle & Ryan 1981) showing a higher rate of predation by bats on calling male frogs *P. pustulosus* than on non-calling males.

These studies all suggest the existence of reproductive costs, and for none of them is it easy to suggest an alternative explanation. Such studies are therefore valuable in revealing the kinds of constraints that can occur. However, correlational studies have been criticized (see, for example, Partridge & Farquhar 1983; Reznick 1985; Bell & Koufopanou 1986; van Noordwijk & de Jong 1986; Partridge & Harvey 1985, 1988; Partridge 1989*a*, *b*). The main reason is that confounding variables are likely to reduce the true magnitude of the effect or even to produce spurious correlations, for example positive correlations when the true values are negative. Individuals can differ in phenotype as a result of their environment in early life or in adulthood. For instance, opportunities for feeding during growth may affect adult body size, identity of parents may affect the social status of offspring, and recent feeding may affect fat reserves of adults. This sort of variation can cause individuals to differ in the size of their options sets; some individuals will then have the potential for high scores on all life-history variables. The environments occupied by different individuals may also differ, affecting both current reproductive success and subsequent survival and fecundity. Both individual phenotype and environment may also differ between populations, and will tend to produce positive correlations between life-history variables.

There have been several surveys of the effect of using correlational studies to deduce options sets (see, for example, Reznick 1985; Bell & Koufopanou 1986; Partridge 1989*a*, *b*). The general picture to emerge is inconsistent, with many positive correlations between life-history characters, which individually lead to high fitness, suggesting that individual phenotype and environmental differences may have a potent influence on the form of individual life histories. This is in itself both a phenomenon of interest and a problem for life-history theory (Partridge 1989*a*).

It is no coincidence that natural variation between individuals has been used to search for reproductive costs, but not for other forms of functional constraint on the life history. It would not be possible to use this technique to look for an association between juvenile survival probability and subsequent adult performance, because once the creature is dead it can yield no data on later parts of its life history. Characters costly in terms of earlier survival must therefore be investigated by other means.

(*c*) *Comparisons of populations or taxa*

The comparative approach is one way of detecting adaptation. If an independent character, such as haplodiploidy, causes a predisposition for the evolution of a dependent character, such as eusociality, then they would be expected to co-occur. The technique has been highly successful in revealing patterns in nature (see, for example, Harvey & Pagel 1991).

In principle, the comparative approach might be used as well to detect constraints on adaptation as to detect adaptation itself, and hence could be useful for measuring options sets. For life histories, species comparisons have been used to show that the large males of highly polygynous species suffer a higher death rate before reaching adulthood. The association probably occurs because males achieve their size excess by rapid growth, at the expense of fat deposition, increasing the chance of starvation (Clutton-Brock *et al.* 1985). Unlike individual variation, the units compared are populations, and so the comparative method can be used to examine functional constraints involving early death. Comparative work has also been useful in studying the trade-off between number and size of offspring (Godfray & Parker, this symposium; Harvey & Keymer, this symposium).

There are difficulties with using the comparative approach to measure options sets. The problem of confounding variables producing spurious correlations, mentioned above in the context of the use of individual differences, applies equally here. In addition, gene–environment interactions can be a problem where different populations or species are placed in a common environment to which they are not equally well adapted (Partridge 1989*b*); this procedure is likely to take less well adapted species away from the boundary and into the interior of their options sets. This may explain why several comparative studies of different species in the laboratory have revealed positive correlations between life history variables (see Snell & King 1977; Schnebel & Grossfield 1988).

Population dynamics pose an additional complication. An association between high seasonal fecundity and low adult survival rates has been reported across species of lizards (Tinkle 1969), birds (O'Connor 1985; Saether 1988) and mammals (Sutherland *et al.* 1986). However, populations in steady state necessarily have equal birth and death rates, so that higher adult fecundity is inevitably accompanied by increased adult or juvenile mortality (Sutherland *et al.* 1986; Sibly & Calow 1987; Gustaffson & Sutherland 1988). One way out of this difficulty would be to confine the use of the technique to life-history characters that make at most only a small contribution to population dynamics. Male reproductive characters could qualify where males contribute only gametes to their progeny. An interesting association between bright male coloration (and hence high reproductive rate?) and short adult lifespan has been reported in lizards (Tinkle 1969), and in general, semelparity seems to be associated with extreme development of male secondary sexual characters (Partridge & Endler 1987), perhaps showing a trade-off between fecundity and longevity for males.

The comparative method was devised to study adaptation, but it is always possible that correlated life-history characters could be independent adaptations to the different study environments. Their association would then not imply any functional constraint. So, for instance, high rates of predation on small individuals might be expected independently to select for speedy development and thus earlier age of first breeding to minimize the time spent in the dangerous size classes, but also to reduce adult reproductive rate as a consequence of the lower value of juveniles. An association between these two sets of characters would therefore not necessarily show a functional constraint between them.

(*d*) *Genetic studies*

Genetic approaches to measuring functional constraints estimate the extent to which genetic variation affecting one life-history character also affects another, measured as the genetic correlation between them. Genetic correlations can be estimated from correlated responses to selection, as in the *Drosophila* example already mentioned (p. 5), or in two-generation studies. An example of the latter is provided by a study of cowpea weevils *Callosobruchus maculatus*, which examined some life-history characters of the female progeny of single males mated to a series of different females. The results showed a negative genetic correlation between adult fecundity and development rate, consistent with the existence of a trade-off constraining the animal's evolutionary options in the culture environment (Moller *et al.* 1989*a*, *b*).

Genetic studies avoid the pitfalls so far mentioned, and they allow all parts of the life history to be studied. It has indeed been argued that only the demonstration of a negative genetic correlation between two traits, caused by pleiotropy, is sufficient to show that there is a functional constraint between them (see, for example, Charlesworth 1980; Lande 1982; Reznick 1985). From this perspective options sets are viewed as a result of genetic constraints imposed by pleiotropy. It is of course true that if, for instance, nutrients are limiting for reproduction and growth, then a genetic variant, the primary mode of action of which is to cause the diversion of nutrients towards reproduction, will also lower growth. To view this as a genetic constraint is perhaps a little perverse; the primary level of action of constraint in the example just described is nutritional, not genetic.

The main argument advanced for the necessity of genetic studies is that, because we are interested in evolutionary options, we need a demonstration that these can be genetically coded. This could be provided by a selection experiment, or by the measurement of genetic correlation (or more generally, the genetic variance-covariance matrix). However, it is not clear that either approach will necessarily reveal the options set. The *reductio ad absurdum* of that position would be to assert that because cheetahs, say, appear to lack any significant genetic variance, their life histories cannot involve trade-offs. Even for persistently large populations, genetic variances give a guide only to the very short term evolutionary future. To find out what could be achieved in the long term in response to a genetic change in the value of one life-history character, two further items of information are needed. The first is the range of effects of new mutational variance. Second, if the initial correlated responses are not the best possible ones, there may be selection on other loci, not initially involved in the response, to modify them. In short, the level of genetic variance available on a reasonable timescale allows exploration of only a very restricted part of the options set.

There is also a pragmatic point, which is that genetic techniques will allow only very slow progress. They confine attention to a small subset of the organisms; it is no coincidence that most genetic studies have been done on insects in the laboratory. The genetic techniques which stand the most chance of being used in the field are selection experiments (see Reznick *et al.* 1990) and two-generation studies. The latter, however, requires the use of controlled breeding designs in an appropriate environment (Clark 1987). There are also problems with large standard errors on estimates of genetic correlation. Even in the carefully controlled laboratory study of weevils described above, in which 761 individuals were scored, several of the estimates of genetic correlation lay outside the theoretical limits of minus and plus one. Such problems frequently occur, and this means that impractically large breeding designs are often required. For this reason perhaps, and because selection magnifies genetic effects (Pease & Bull 1988), correlated responses to selection have provided more evidence for the existence of genetic constraint than has the two generation approach (Reznick 1985; Lessells 1991). However, both approaches are limited by the amount of genetic variation available, and genetic studies are therefore probably not going to reveal why an albatross lays one egg and a blue tit ten.

These sorts of considerations provide a strong motivation for finding other avenues of approach. The main alternative avenue is experimental manipulation. However, use of this approach involves assumptions, which require empirical validation.

(*e*) *Experimental manipulations*

Experimental manipulations examine the effect of manipulation of one life-history variable on others. For instance, the cost of reproduction in female fruitflies *D. subobscura* was investigated by exposing them briefly to X-irradiation or high temperatures, both of which had the apparently paradoxical effect of extending the lifespan. There was, however, no significant effect on mutant ovaryless females, suggesting that the manipulations abolished ovarian activity, and that ovarian activity reduced longevity (Maynard Smith 1958; Lamb 1964). The geographical strain of flies used in the experiments mated only once. In the related *D. melanogaster* remating is common, and here mating reduces female lifespan even more than does the production of eggs (Fowler & Partridge 1989).

Surveys of the results of experiments have shown that, like correlated responses to selection, they reveal fairly consistent evidence for the existence of functional constraints (Reznick 1985; Bell & Koufopanou 1986; Partridge 1989*a*). Could experiments of this general kind be used to measure options sets? The approach has potential strengths and weaknesses.

A strength of experiments is that they have the potential for exploration of a large range of character values, much greater than could be achieved in practice for most organisms by artificial selection. In addition, experiments are strictly prospective; they will reveal only effects occurring after the point in the life history at which the intervention is made. This allows precise examination of the present and future effects of decisions at each point in the life history in isolation from others but, on the other hand, would not necessarily reveal the most likely evolutionary response to a change in the manipulated life-history variable, because the correlated response could involve an event earlier in the life history. For instance, male swallows *Hirundo rustica* with experimentally elongated tails had higher reproductive success (Moller 1988) but incurred a cost in terms of their reproductive success in the ensuing breeding season (Moller 1989). However, there could be additional earlier survival costs for a bird growing a longer tail. Experiments have so far been used mainly to study costs of reproduction, for which some manipulative techniques are available. Ignorance of the machinery controlling developmental decisions has impeded manipulation of earlier parts of the life history.

Experiments are not free of pitfalls. One is that the manipulation used may alter the options set. For instance, using food to induce a female fruitfly to lay more eggs or a male swallow to grow a longer tail could also have a direct effect on the dependent variables survival and future reproduction. This is a serious difficulty, which can only be circumvented by an understanding of physiological control mechanisms. Manipulations should ideally alter all aspects of the target variable in a coordinated way. For instance, some brood size manipulation experiments, aimed at investigating the cost of chick-rearing, have revealed that the parents do not alter their rate of feeding the brood in response to a change in its size (reviewed in Lessells 1991). Even in simpler systems there can be pitfalls. Abolition of ovarian activity increased the lifespan of female *D. subobscura*, but these experiments may not have fully abolished reproductive costs, because the fat body synthesis and degradation of yolk polypeptides may have continued in the absence of an ovary.

If experiments are to be used successfully to measure options sets, then the mechanisms underlying the response to the manipulation must be studied; an understanding of how the organism controls its own life-history events is needed for effective intervention. This problem is especially acute if we want to understand the effects of decisions made in the juvenile period on events then and in adulthood. As a result, we really have very little idea of how much adult survival and fecundity are determined by decisions made in adulthood and how much on ones made earlier in life.

If appropriate experimental manipulations can be found, will they help us to understand life-history evolution? For them to do so, two things would have to be true. The system would have to give its best response to the manipulation, which is tantamount to saying that the relevant phenotypic plasticity would have to be optimal. It would also have to have the same characteristics as the probable evolutionary response to a genetic change of the same magnitude. These are empirically testable assumptions, and there is a need for studies of the relation between responses to experimental manipulations and correlated response to selection. If the two methods give the same results, then the genetical relevance of the experimental manipulations has been shown. Otherwise two types of differences are possible. Experimental manipulation could show options not revealed by a genetical study, and this is the likely outcome if there is an insufficiency of genetic variation for selection to act on. Alternatively, the genetical study might show options not revealed by experimental manipulation, if genes can achieve more than the experimental manipulator. It may be that evolutionary responses and correlated responses are achieved by a multiplicity of intermediate mechanisms that cannot be reproduced without going through the evolutionary process itself; it would certainly be worth trying to find out.

Thus, no one approach to the elucidation of options sets is likely to be sufficient on its own. There will always be a place for individual studies that reveal correlations hard to explain except in terms of trade-offs. If these trade-offs are real, it should be possible to show them using genetic techniques, and where this can be achieved it is the method of choice, since it is certain that all the options revealed are genetically codable. The major problems with the genetic approach are that it may not reveal all possible genetic options, and that it is limited by the amount of genetic variation present in the study populations. These problems may be to some extent circumvented using experimental manipulation, successful application of which depends on a secure understanding of the physiological mechanisms involved in the control of life-history variables. However, an indication is needed that what was achieved by experimental manipulation could also be achieved by genetic engineering. This may not be implausible, however. For instance, in the experiment of Fowler & Partridge (1989) discussed above, it was shown that in an environment containing two males per female, reducing mating by $\frac{2}{3}$ improved female survivorship. It seems clear that in that environment, genes that achieved that reduction in mating would obtain the survivorship advantage. In principle, reduction in mating could be achieved by many mechanisms, although whether these could all be achieved in the study environment is less sure.

5. LONG-TERM CONSTRAINTS

Variation in the shapes of trade-off curves and differences in selection pressures may not be a sufficient explanation for life-history diversity. There are many instances where organisms with otherwise similar biology and inhabiting similar environments show marked differences in their life histories.

Part of the explanation could be provided by trade-off curves themselves if they are complex in shape, and hence allow more than one locally optimum solution. For instance, among the Salmonidae, the Genus *Salmo* is iteroparous, while *Oncorhynchus* is nearly always semelparous. These two genera coexist, spawning and maturing in the same areas. Schaffer (1974) suggested that the difference might be caused by a complex trade-off curve between fecundity and subsequent survival, with optima at both semelparity and iteroparity. Under these circumstances the point on the trade-off curve initially occupied could be important in determining the response of the life history to natural selection.

Even with multiple optima, the shapes of trade-off curves will be informative only about the short-term courses available. Over evolutionary time, options sets themselves can change. The extent to which they can do so will depend upon whether the underlying constraint is universal or local (Maynard Smith *et al.* 1985). Local constraints, which apply only to specific taxa, will mean that differences in design between organisms will affect the way that they evolve in a common environment. For instance, pre-existing contact with offspring appears to make it easier for fish to evolve parental care (Gross & Shine 1981).

Evolutionary lags may be another source of variation in life histories. Time is necessary both for the possible mutational variance to occur, and for a response to selection to be effected, so that the observed phenotype may lag behind the optimal one during periods of evolution, especially if selection is weak. Evolutionary delays of this kind could both withhold a population from the optimum, and cause diversity among populations subject to similar selection pressures for different lengths of time. The history of a population will then be a determinant of its life history.

Under some circumstances historical effects can take the extreme form of irreversibility, so that organisms can become stuck in absorbing states (Bull & Charnov 1985). One reason is that some evolutionary paths may involve loss of structures or processes which are then difficult to re-evolve. For instance, it could be easy to evolve semelparity from iteroparity by loss of late reproduction, but more difficult to evolve in the reverse direction, if the genetic substrate for post-reproductive survival has been lost. Natural selection itself can also cause irreversibility. For instance, in parasitic wasps, clutch size is bimodal, with clutch sizes of one giving rise to larvae equipped with large mandibles that are used to kill any introduced larvae, and clutch sizes above three giving rise to gregarious non-armed larvae.

This situation seems to be a consequence of parent–offspring conflict rather than a reflection of the distribution of parental optima. A theoretical analysis (Godfray 1987) suggested that a rare gene for gregariousness could only with difficulty invade a population of fighting larvae unless fitness increases with clutch size. The reason is that a rare gregarious larva is likely to find itself with at least one fighter, and will hence perish. Under these circumstances it pays the parent to reduce the clutch size to one, because only one larva will give rise to an adult, and clutch size one becomes an absorbing state. Evolving fighting larvae from gregarious ones would not present the same difficulty.

These kinds of long-term constraints are not easy to investigate. Theoretical studies like the one just described are likely to prove extremely valuable, but so also are empirical studies of trade-off curves and of the life histories, of populations with well known evolutionary histories.

REFERENCES

Bell, G. & Koufopanou, V. 1986 The cost of reproduction. In *Oxford surveys in evolutionary biology*, vol. 3 (ed. R. Dawkins & M. Ridley), pp. 83–131. Oxford University Press.

Bull, J. J. & Charnov, E. L. 1985 On irreversible evolution. *Evolution* **39**, 1149–1155.

Calow, P. 1979 The cost of reproduction: a physiological approach. *Biol. Rev.* **54**, 23–40.

Calow, P. & Sibly, R. M. 1990 A physiological basis of population processes: ecotoxicological implications. *Funct. Ecol.* **4**, 283–288.

Caswell, H. 1989 Life history strategies. In *Ecological concepts* (ed. J. M. Cherrett), pp. 285–307. Oxford: Blackwell Scientific Publications.

Charlesworth, B. 1980 *Evolution in age-structured populations*. Cambridge University Press.

Charlesworth, B. 1990 Optimization models, quantitative genetics and mutation. *Evolution* **44**, 520–538.

Charlesworth, B. & Leon, J. A. 1976 The relation of reproductive effort to age. *Am. Nat.* **110**, 449–459.

Charnov, E. L. 1989 Phenotypic evolution under Fisher's Fundamental Theorem of Natural Selection. *Heredity* **62**, 113–116.

Charnov, E. L. & Krebs, J. R. 1974 On clutch size and fitness. *Ibis* **116**, 217–219.

Clark, A. G. 1987 Genetic correlations: the quantitative genetics of evolutionary constraints. In *Genetic constraints on adaptive evolution* (ed. V. Loeschcke), pp. 25–45. Berlin: Springer-Verlag.

Clutton-Brock, T. H., Albon, S. D. & Guiness, F. E. 1985 Parental investment and sex differences in juvenile mortality in birds and mammals. *Nature, Lond.* **313**, 131–133.

Clutton-Brock, T. H., Guiness, F. E. & Albon, S. G. 1982 *Red deer – behaviour and ecology of two sexes*. University of Chicago Press.

Cody, M. L. 1966 A general theory of clutch size. *Evolution* **20**, 174–184.

Cole, L. C. 1954 The population consequences of life history phenomena. *Q. Rev. Biol.* **29**, 103–137.

Dhondt, A. A., Adriaensen, F., Matthysen, E. & Kempenaers, B. 1991 Non-adaptive clutch sizes in tits: evidence for the gene flow hypothesis. *Nature, Lond.* **348**, 723–725.

Fisher, R. A. 1958 *The genetical theory of natural selection*. New York: Dover.

Fowler, K. & Partridge, L. 1989 A cost of mating in female fruitflies. *Nature, Lond.* **338**, 760–761.

Gadgil, M. & Bossert, W. H. 1970 Life historical consequences of natural selection. *Am. Nat.* **104**, 1–24.

Godfray, H. C. J. 1987 The evolution of clutch size in parasitic wasps. *Am. Nat.* **129**, 221–233.

Gould, S. J. & Lewontin, R. C. 1979 The spandrels of San Marco and the Panglossian paradigm: a critique of the adaptationist programme. *Proc. R. Soc. Lond.* B **205**, 581–598.

Gross, M. R. 1985 Disruptive selection for alternative life histories in salmon. *Nature, Lond.* **313**, 47–48.

Gross, M. R. & Shine, R. 1981 Parental care and mode of fertilization in ectothermic vertebrates. *Evolution* **35**, 775–793.

Gustaffson, L. 1986 Lifetime reproductive success and heritability: empirical support for Fisher's fundamental theorem. *Am. Nat.* **128**, 761–764.

Gustaffson, L. & Part, T. 1990 Acceleration of senescence in the collared flycatcher *Ficedula albicollis* by reproductive costs. *Nature, Lond.* **347**, 279–281.

Gustaffson, L. & Sutherland, W. J. 1988 The costs of reproduction in the collared flycatcher *Ficedula albicollis*. *Nature, Lond.* **335**, 813–815.

Harvey, P. H. & Pagel, M. D. 1991 *The comparative method*. Oxford University Press.

Hirshfield, M. F. & Tinkle, D. W. 1975 Natural selection and the evolution of reproductive effort. *Proc. natn. Acad. Sci. U.S.A.* **72**, 2227–2231.

Lack, D. 1954 *The natural regulation of animal numbers*. Oxford University Press.

Lamb, M. J. 1964 The effects of radiation on the longevity of female *Drosophila subobscura*. *J. Insect Physiol.* **10**, 487–497.

Lande, R. 1982 A quantitative genetic theory of life history evolution. *Ecology* **63**, 609–615.

Law, R. 1979 Ecological determinants in the evolution of life histories. In *Population dynamics* (ed. R. M. Anderson, B. D. Turner & L. R. Taylor), pp. 81–103. Oxford: Blackwell Scientific Publications.

Lessells, C. M. 1991 The evolution of life histories. In *Behavioural ecology: an evolutionary approach*, 3rd edn. (ed. J. R. Krebs & N. B. Davies). Oxford: Blackwell Scientific Publications. (In the press.)

Levins, R. 1968 *Evolution in changing environments*. Princeton University Press.

Lewontin, R. C. 1965 Selection for colonising ability. In *The genetics of colonising species* (ed. H. G. Baker & G. Ledyard Stebbins), pp. 77–94. New York: Academic Press.

Linden, M. & Moller, A. P. 1989 Cost of reproduction and covariation of life history traits in birds. *Trends Ecol. Evol.* **4**, 367–371.

Maynard Smith, J. 1958 The effects of temperature and of egg-laying on the longevity of female *Drosophila subobscura*. *J. exp. Biol.* **35**, 832–842.

Maynard Smith, J. 1978 Optimization theory in evolution. *A. Rev. Ecol. Syst.* **9**, 31–56.

Maynard Smith, J. 1982 *Evolution and the theory of games*. Cambridge University Press.

Maynard Smith, J., Burian, R., Kauffman, S., Alberch, P., Campbell, J., Goodwin, B., Lande, R., Raup, D. & Wolpert, L. 1985 Developmental constraints and evolution. *Q. Rev. Biol.* **60**, 265–287.

Moller, A. P. 1988 Female choice selects for male sexual tail ornaments in the monogamous swallow. *Nature, Lond.* **332**, 640–642.

Moller, A. P. 1989 Viability costs of male tail ornaments in a swallow. *Nature, Lond.* **339**, 132–135.

Moller, H., Smith, R. H. & Sibly, R. M. 1989 Evolutionary demography of a bruchid beetle. I. Quantitative genetical analysis of the female life history. *Funct. Ecol.* **3**, 673–681.

Moller, H., Smith, R. H. & Sibly, R. M. 1989 Evolutionary demography of a bruchid beetle. II. Physiological manipulations. *Funct. Ecol.* **3**, 683–691.

Mueller, L. D. & Ayala, F. J. 1981 Trade-off between *r*-selection and *K*-selection in Drosophila populations. *Proc. natn. Acad. Sci. U.S.A.* **78**, 1303–1305.

O'Connor, R. J. 1985 Behavioural regulation of bird populations: a review of habitat use in relation to

migration and residency. In *Behavioural ecology: ecological consequences of adaptive behaviour* (ed. R. M. Sibly & R. H. Smith), pp. 105–142. Oxford: Blackwell Scientific Publications.
Orzack, S. H. & Tuljapurkar, S. 1989 Population dynamics in variable environments. VII. The demography and evolution of iteroparity. *Am. Nat.* **133**, 901–923.
Parker, G. A. & Maynard Smith, J. 1990 Optimality theory in evolutionary biology. *Nature, Lond.* **348**, 27–33.
Partridge, L. 1989*a* Lifetime reproductive success and life history evolution. In *Lifetime reproduction in birds* (ed. I. Newton), pp. 421–440. London: Academic Press.
Partridge, L. 1989*b* An experimentalist's approach to the role of costs of reproduction in the evolution of life histories. In *Toward a more exact ecology* (ed. P. J. Grubb & J. B. Whittaker), pp. 231–246. Oxford: Blackwell Scientific Publications.
Partridge, L. & Endler, J. 1987 Life history constraints on sexual selection. In *Sexual selection: testing the alternatives* (ed. J. Bradbury & M. Andersson), pp. 265–277. Chichester: Wiley.
Partridge, L. & Farquhar, M. 1983 Lifetime mating success of male fruitflies *Drosophila melanogaster* is related to their size. *Anim. Behav.* **31**, 871–877.
Partridge, L. & Harvey, P. H. 1985 Costs of reproduction. *Nature, Lond.* **316**, 20–21.
Partridge, L. & Harvey, P. H. 1988 The ecological context of life history evolution. *Science, Wash.* **241**, 1449–1454.
Pease, C. M. & Bull, J. J. 1988 A critique of methods for measuring life history trade-offs. *J. evol. Biol.* **1**, 293–303.
Pettifor, R. A., Perrins, C. M. & McCleery, R. H. 1988 Individual optimization of clutch size in great tits. *Nature, Lond.* **336**, 160–162.
Pianka, E. R. & Parker, W. S. 1975 Age-specific reproductive tactics. *Am. Nat.* **109**, 453–464.
Price, T. R., Kirkpatrick, M. & Arnold, S. J. 1988 Directional selection and the evolution of breeding date in birds. *Science, Wash.* **240**, 798–799.
Reznick, D. 1985 Costs of reproduction: an evaluation of the empirical evidence. *Oikos* **44**, 257–267.
Reznick, D. A., Bryga, H. & Endler, J. A. 1990 Experimentally induced life history variation in a natural population. *Nature, Lond.* **346**, 357–359.
Rose, M. 1984 Laboratory evolution of postponed senescence in *Drosophila melanogaster*. *Evolution* **38**, 1004–1010.
Rose, M. R., Service, P. M. & Hutchinson, E. W. 1987 Three approaches to trade-offs in life history evolution. In *Genetic constraints on adaptive evolution* (ed. V. Loeschcke), pp. 91–105. Berlin: Springer-Verlag.
Saether, B-E. 1988 Pattern of covariation between life history traits of european birds. *Nature, Lond.* **331**, 616–617.
Schaffer, W. M. 1974 Selection for life histories: the effects of age structure. *Ecology* **55**, 291–303.
Sibly, R. M. 1989 What evolution maximises. *Funct. Ecol.* **3**, 129–135.
Sibly, R. M. 1991 The life-history approach to physiological ecology. *Funct. Ecol.* (In the press.)
Sibly, R. M. & Calow, P. 1986 *Physiological ecology of animals*. Oxford: Blackwell Scientific Publications.
Sibly, R. M. & Calow, P. 1987 Ecological compensation – a complication for testing life history theory. *J. theor. Biol.* **125**, 177–186.
Snell, T. E. & King, C. E. 1977 Life span and fecundity patterns in rotifers: the cost of reproduction. *Evolution* **31**, 882–890.
Schnebel, E. M. & Grossfield, J. 1988 Antagonistic pleiotropy: an interspecific *Drosophila* comparison. *Evolution* **42**, 306–311.
Stearns, S. C. 1976 Life-history tactics: a review of the ideas. *Q. Rev. Biol.* **51**, 3–47.
Stearns, S. C. 1989 Trade-offs in life history evolution. *Funct. Ecol.* **3**, 259–268.
Sutherland, W. J., Grafen, A. & Harvey, P. H. 1986 Life history correlations and demography. *Nature, Lond.* **320**, 88.
Tinkle, D. W. 1969 The concept of reproductive effort and its relation to the evolution of life histories of lizards. *Am. Nat.* **103**, 501–516.
Tuttle, M. D. & Ryan, M. J. 1981 Bat predation and the evolution of frog vocalizations in the neotropics. *Science, Wash.* **214**, 677–678.
van Noordwijk, A. J. & de Jong, G. 1986 Acquisition and allocation of resources: their influence of variation in life history tactics. *Am. Nat.* **128**, 137–142.
van Noordwijk, A. J., van Balen, J. H. & Scharloo, W. 1980 Heritability of ecologically important traits in the great tit. *Ardea* **68**, 193–203.
Williams, G. C. 1966 Natural selection, the costs of reproduction, and a refinement of Lack's principle. *Am. Nat.* **100**, 687–690.

Discussion

R. J. H. Beverton (*Montana, Old Roman Road, Langstone, Gwent NP6 2JU, U.K.*). Although commending your comparative approach to life-history dynamics, care is needed about the extent to which quantitative relations such as that between pre-mature mortality rate and subsequent spawning strategy (semelparity against iteroparity) are valid between different animal groups. Perhaps I could give three examples from fish to illustrate this. Populations of American shad (*A. shapidissima*) at the southern end of their range are semelparous and have a high fecundity, whereas at the northern end they are predominantly iteroparous and have a low fecundity, which fits your prognosis. However, the Pacific pink salmon (*O. gorbuscha*) is strictly semelparous, spawning and dying at two years of age, whereas the Atlantic salmon (*S. salar*) is iteroparous, being able (if allowed to survive) to spawn a number of times; yet both species have a similar fecundity and pre-mature mortality rate. Lastly, elasmobranchs produce a very few young each year and have a low pre-mature mortality rate; whereas many teleosts are highly fecund, laying up to a million or so eggs per adult annually, and have a correspondingly high pre-mature mortality rate. Yet iteroparity is common in both groups. Clearly, fecundity and pre-mature mortality rate must vary inversely for the population to be balanced, and different kinds of fish have solved that part of the life-history equation in quite different ways independently of subsequent spawning strategy.

L. Partridge and R. Sibly. As we have suggested, great care is needed in applying the comparative approach. When different species are compared, it is necessary first to consider whether they have the same options sets – in general we imagine the answer is no. Complexity of the shape of the options sets could also produce multiple optima. But with no information about the shape of the options sets, no predictions can be made from optimality theory. For the elasmobranch and teleost comparison you mention, there is likely to be a trade-off between egg size and number, which would explain the fecundity difference, and which would not be expected to influence the degree of iteroparity.

W. G. Hill (*University of Edinburgh, Edinburgh, U.K.*). Different species adopt widely different reproductive strategies in what appear quite similar environments, for example, elasmobranchs and teleosts as mentioned previously. This would suggest that organisms can utilize their energy and other resources effectively in many ways. Further, it suggests that curves relating total fitness to any single component of fitness, providing all other components are altered appropriately, are very flat: there is no, or a very weak, global optimum. However, a curve relating fitness to any component on its own, or any pair or more of components with others ignored, would not be flat. Thus, I am concerned as to what, if any, conclusions one can obtain from studies on subsets of traits, or am I missing something important?

It seems to me that an experimental approach, albeit not necessarily a practical one, to answering the question of whether there is a global optimum strategy, i.e. a fitness maximum, and how pronounced it is, would be to maintain in a constant laboratory environment two or more species with very different current strategies and observe their evolution. Convergence should be expected, but at what rate?

L. Partridge and R. Sibly. We fully accept that a multidimensional approach is desirable, indeed essential for a full understanding. However, even if the fitness hypersurface is flat in the n-dimensional space with axes the life-history characters, it does not follow that the trade-off hypersurface, bounding a species' options set, will also be flat. Thus, the elasmobranchs and the teleosts probably have very different options sets, owing in part to their different body designs.

Evolution of senescence: late survival sacrificed for reproduction

T. B. L. KIRKWOOD[1] AND M. R. ROSE[2]

[1] *National Institute for Medical Research, Mill Hill, London NW7 1AA, U.K.*
[2] *Department of Ecology and Evolutionary Biology, University of California, Irvine, California 92717, U.S.A.*

SUMMARY

In so far as it is associated with declining fertility and increasing mortality, senescence is directly detrimental to reproductive success. Natural selection should therefore act in the direction of postponing or eliminating senescence from the life history. The widespread occurrence of senescence is explained by observing that (i) the force of natural selection is generally weaker at late ages than at early ages, and (ii) the acquisition of greater longevity usually involves some cost. Two convergent theories are the 'antagonistic pleiotropy' theory, based in population genetics, and the 'disposable soma' theory, based in physiological ecology. The antagonistic pleiotropy theory proposes that certain alleles that are favoured because of beneficial early effects also have deleterious later effects. The disposable soma theory suggests that because of the competing demands of reproduction less effort is invested in the maintenance of somatic tissues than is necessary for indefinite survival.

1. INTRODUCTION

Senescence is a standard feature in the life histories of higher animals (Comfort 1979). It is usually defined in relation to the pattern of age-specific mortality, a population being said to experience senescence if it exhibits a progressive increase in the age-specific death rate even when the population is maintained under conditions that are ideal for survival (see, for example, Medawar 1955; Maynard Smith 1962). Underlying this progressive increase in the age-specific death rate is a generalized deterioration in a broad spectrum of physiological and metabolic functions (Finch & Schneider 1985). These physiological decrements leave the organism increasingly vulnerable to a variety of intrinsic and extrinsic factors that may cause death. An important correlate of the declining physiological competence of a senescing organism is that reproduction also generally declines with age. Taken together, the declines in survival and fecundity mean that a senescing organism experiences a major, and eventually total, loss in fitness during later ages. The puzzle, therefore, is to explain why this trait, which is deleterious to individual fitness, has evolved.

Senescence is most clearly seen in the case of a species with an iteroparous life history (Kirkwood 1985). In the iteroparous life history, the adult is capable of repeated reproduction after gaining sexual maturity (Cole 1954). This life-history pattern is potentially open-ended: it could in principle extend indefinitely, if senescence did not bring it to a close. This is in contrast to the semelparous life history, where death tends to follow closely upon reproduction, often as a direct result of endocrine and other changes which accompany the physiological commitment to reproduce (see, for example, Robertson 1961; Wodinsky 1977). Senescence is also most clearly seen in species where there is a clear distinction between germ-line and somatic tissue. When this distinction is lacking, organisms tend to be capable of reproducing vegetatively. Vegetative reproduction blurs the concept of individual survivorship and makes it harder to define whether or not senescence occurs. In fact, species capable of vegetative reproduction provide the best examples of organisms that do not senesce (Comfort 1979; Bell 1984).

Why does senescence occur and what determines its rate of progress? These questions require answers at both the proximate, physiological level and at the ultimate, evolutionary level. The field of gerontology is replete with physiological theories to account for senescence (Finch & Schneider 1985; Warner *et al.* 1987; Medvedev 1990). Evolutionary theories explain senescence in terms of the selection forces acting on the life history. There is a long tradition of evolutionary discussion of senescence, which began last century with the work of Weismann and Wallace (Kirkwood & Cremer 1982; Rose 1991).

In this paper, we describe two explanations of senescence which converge in the suggestion that its evolution can best be understood as a by-product of the priority that natural selection places on reproduction. One explanation, based in population genetics and owing mainly to Williams (1957), is the theory of 'antagonistic pleiotropy' (see also Charlesworth 1980; Rose 1984*b*). The other explanation is the 'disposable soma' theory (Kirkwood 1977, 1981; Kirkwood & Holliday 1979). The disposable soma theory is based on an optimality approach, consistent with the framework of physiological ecology (Townsend & Calow 1981). The common element in both these theories is the conclusion that, in effect, natural selection trades late survival for enhanced early fecundity. The theories differ in the extent to which they also address the question of physiological mechanisms of senescence (see also Discussion).

2. EVOLUTION OF SENESCENCE

(*a*) *Population genetics*

The first hints of a population genetic approach to the problem of senescence come from Fisher (1930, pp. 28–29) and Haldane (1941, pp. 192–194). Both Fisher and Haldane suggested that the force of natural selection acting on allelic variants affecting survival should decline during adulthood. Hamilton (1966) and Charlesworth (1980) later showed mathematically that this intuition is often correct: under conditions where the Malthusian parameter defines fitness, the intensity of selection acting on an allele modifying survival probability by a fixed proportion will decline with age. That is, the force of natural selection acting on adult survival does indeed tend to decline with adult age. Comparable results also apply to selection acting on age-specific fertility, although the period of decline may be shifted (Hamilton 1966; Charlesworth 1980). Thus it may be generally concluded that the action of natural selection on age-specific fitness effects declines with age.

Given that natural selection is the force responsible for the adaptation of the organism, the evolutionary principle that the action of natural selection declines with age during the adult stage of the life cycle leads to the prediction that senescence should evolve (Medawar 1946, 1952; Williams 1957).

One population-genetic mechanism through which senescence might evolve is the accumulation of deleterious mutations that only act later in life, when the action of natural selection is extremely weak. This idea was discussed extensively by Medawar (1952). Even if senescence does not exist already in the life history, the lifespan of most animals is effectively curtailed by the mortality exacted by the environment. This provides the scope for late-acting deleterious mutations to accumulate relatively immune to the action of natural selection. When a rare individual lives long enough to encounter the effects of these late-acting mutations, they combine to generate the diverse pathologies of the aged adult. Thus, mutations accumulate and introduce senescence into the life history.

A variant of the mutation accumulation idea, also discussed by Medawar (1952), is that natural selection acts on alleles at age-of-action modifier loci to postpone numerous genetic diseases from earlier to later ages. However, the magnitude of the selection pressure for postponement of genetic diseases is only on the order of the mutation rate (cf. Ewens 1979, pp. 195–198). For this reason, it is unlikely that such selection will overcome mutation pressure acting on the modifier locus, and this population genetic mechanism for the evolution of senescence is not given credence (Charlesworth 1980, p. 219).

Mutation accumulation is essentially a neutral process reflecting the inability of selection to exert tight control over the later portion of the lifespan. A stronger theory is obtained if it is assumed that the late deleterious effects are the pleiotropic consequences of genes that are favoured by selection because they confer early fitness benefits (Williams 1957). This is the theory of antagonistic pleiotropy.

Antagonistic pleiotropy introduces the idea of a trade-off between early benefit and late cost, and the important thing about the declining action of selection with adult age is that it takes only a small fitness benefit early in life to outweigh a substantial deleterious effect later on. An interesting technical point is that the presence of antagonistic pleiotropy may maintain variability in relation to age-specific fitness effects (Rose 1982, 1985). This will then make it difficult to detect the action of mutation accumulation when antagonistic pleiotropy has also been involved in the evolution of senescence. However, while the action of one of these population genetic mechanisms may impede the detection of the other, there is no theoretical difficulty with their simultaneous action in one species.

(*b*) *Physiological ecology*

The disposable soma theory for the evolution of senescence comes from looking at the problem physiologically and asking how an organism should best allocate resources among the various metabolic tasks it needs to perform (Kirkwood 1977, 1981; Kirkwood & Holliday 1979, 1986). In particular, the disposable soma theory addresses the question of the optimal investment of resources in somatic maintenance.

In physiological terms, an organism is an entity that takes in resources from its environment, primarily energy in the form of nutrients, uses these resources for a variety of metabolic tasks such as growth and maintenance, and in due course reproduces to generate an output of progeny. The problem of allocation of resources arises because resources used for one purpose are no longer available for other purposes. A central issue in physiological approaches to life-history evolution (see Sibly & Calow 1986; Partridge & Harvey 1988) is to understand which of the many different allocation strategies is optimal, i.e. maximizes fitness, for an organism subject to natural selection under a given set of ecological constraints. We might note that there is no implicit assumption here that resources are necessarily scarce. In practice, resources are often limiting. However, even where resources are abundant there are constraints on how fast they can be utilized. The point is that no matter what the gross intake of resources, there is always the problem of how best to divide them.

General solutions to the problem of optimal resource utilization are elusive, chiefly because the constraints that determine the option sets are as yet unknown (Partridge & Sibly, this symposium). Nevertheless, informative studies of specific trade-offs can be made. Senescence can be understood by focusing on the investment in maintenance of somatic, i.e. non-

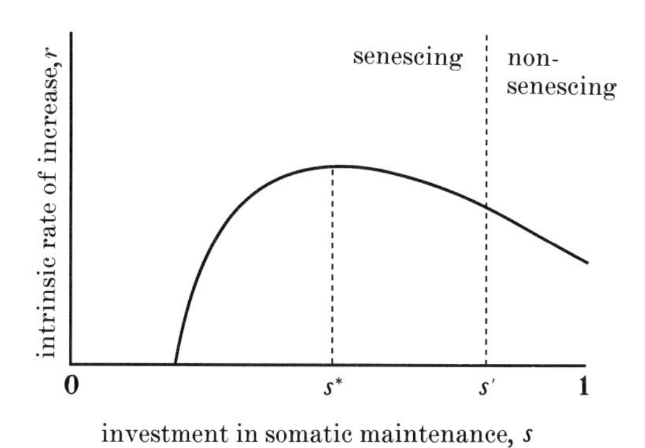

Figure 1. Relation between fitness, measured as the intrinsic rate of increase r, and the level of investment in somatic maintenance s. For $s > s'$, senescence does not occur. The optimum $s = s^*$ is predicted to be always less than s'.

reproductive, parts of the body. We ask two questions. First, for any given iteroparous species is there a best level of investment in maintenance? Secondly, is this sufficient to maintain the organism in a steady, non-senescing condition? The second of these questions presupposes that it is in fact possible to maintain the organism well enough for it to survive without progressive deterioration. If this is not the case, then there is nothing to explain about the evolution of senescence, as senescence is inevitable.

Formal ecological models can be used to answer these questions if we define fitness in terms of the Malthusian parameter, r, obtained from solving the standard Lotka equation

$$\int e^{-rx} l(x;s)\, m(x;s)\, dx = 1,$$

where survivorship $l(x;s)$ and fecundity $m(x;s)$ are written as functions depending not only on age, x, but also on the investment in maintenance, s (see, for example, Kirkwood 1990; see Appendix). From such models, we obtain fitness curves relating r to s. The typical form of the fitness curves is shown in figure 1, and it can be seen from figure 1 that an optimum value for s does exist at $s = s^*$, and that s^* is less than the minimum investment in maintenance ($s = s'$) required to be in the non-senescing condition.

This result establishes the conclusion that the optimum level of investment in somatic maintenance for an iteroparous species is less than the level required for indefinite survival. In other words, natural selection should favour a strategy that results in the progressive accumulation of unrepaired somatic defects, which in turn must result in senescence. This conclusion is robust to a wide range of variation in the specific assumptions of the ecological model used to generate figure 1, and can in fact be understood quite well without recourse to a formal model. This can be done by imagining what happens if an organism moves in from either of the ends of the horizontal scale in figure 1.

If nothing is invested in maintenance ($s = 0$), the organism will senesce with great rapidity. It is easy to see that raising the investment in maintenance increases fitness, at least initially. On the other hand, investing maximally in maintenance ($s = 1$) means that the organism is well inside the non-senescing region ($s > s'$). A small reduction in maintenance does not impose a survivorship cost but releases additional resources for growth and reproduction, and so leads to increased fitness. The same argument can be repeated until the boundary between the senescing and non-senescing states is reached. At this point, the organism is investing just enough in maintenance not to senesce, but is nevertheless subject to extrinsic hazards of the environment. If environmental mortality is, say, 50% a year, then the chance of surviving more than 20 years is less than one in a million. For practicable purposes this chance is negligible, so nothing is really lost if the investment in maintenance is reduced a little further so the organism now senesces at age 20 years. Not only is nothing lost in terms of survivorship, but further resources are released for growth and reproduction. This means that fitness continues to rise as the investment in maintenance is further reduced, in other words, as the boundary is crossed from the non-senescing to the senescing condition.

This verbal argument is sufficient to explain the form of the fitness curve in figure 1. Because fitness rises as the organism moves away from either extreme ($s = 0$ or $s = 1$), there is an optimum at some intermediate value. (The argument does not establish the uniqueness of the optimum, but for a simple life history a unique optimum is likely.) Secondly, the optimum occurs at a lower investment in somatic maintenance than is needed for indefinite survival. The optimum occurs at the point where the cost to survival of further reducing the investment in maintenance becomes large enough to cancel the benefit of any resulting increase in the rates of growth and reproduction.

This explanation for the evolution of senescence is termed the disposable soma theory for its analogy with the manufacture of disposable goods. In essence, the theory recognizes that all that is needed is a soma that remains in good condition through its normal expectation of life in the wild. Better maintenance than this is a waste, so the optimum is less than what is required for indefinite survival. This leads to the explicit predictions that (i) senescence is the result of the accumulation of somatic defects, and (ii) longevity is regulated through the efficiency of somatic maintenance processes.

3. EVOLUTION OF LIFESPAN

Species exhibit great variation in their lifespans. Among mammals alone, species' lifespans range over nearly two orders of magnitude. How does the evolution of species' lifespan differences relate to the theories on evolution of senescence?

We have just seen in the disposable soma theory that it is the presence of environmental mortality which makes it not worthwhile to invest in better maintenance than is needed to preserve somatic functions through the normal expectation of life in the wild. Expectation of life in the wild for an iteroparous species is largely determined by the prevailing level of environmental mortality. This is because most deaths occur in young animals through accidents, predation, and infectious

Table 1. *Evolution of lifespan in response to varying the level of environmental mortality*

(The centre row corresponds to a life history fitted to data from the mouse, *Mus musculus* (see Appendix). The minimum maintenance level required for indefinite lifespan (s') is set at 0.8.)

environmental mortality[a]	maximum reproductive rate[b]	optimal maintenance level	lifespan[c]
10.0	8.2	0.37	20
5.0	5.2	0.45	29
3.2	4.0	0.50	36
2.0	3.2	0.56	45
1.0	2.2	0.65	70

[a] % per month.
[b] births female^{-1} month^{-1}.
[c] 99th percentile of adult survival (months).

disease, with many species showing little or no evidence of senescence in the wild (see, for example, Lack 1954; Promislow 1991).

When the level of environmental mortality is high, it is less worthwhile to invest heavily in maintenance and more worthwhile to invest in rapid growth and reproduction, and vice versa. Thus, it is clear that in the disposable soma theory the major driving force in the evolution of longevity is likely to be the prevailing level of environmental mortality. This can be studied in detail with the model described in the Appendix. The model is applied by requiring that the set of parameter values defining the survivorship and fecundity sechedules, $l(x;s)$ and $m(x;s)$, satisfy the constraint $r = 0$ for $s = s^*$, in other words, that a species should be at ecological equilibrium.

Starting from a point in the parameter space which satisfies this constraint, one can explore trajectories through this point which correspond to allowable evolutionary paths. An example is shown in table 1, where the trajectory tracks the effects of varying the level of environmental mortality. The centre row of the table defines a starting point based on data for the mouse, *Mus musculus* (see Appendix). Colonization of a more dangerous niche, equivalent to moving up the table, can occur successfully only if there is a concomitant increase in fecundity. The optimum level of investment in maintenance is reduced, together with lifespan. Moving in the other direction, down the table, a reduction in the level of environmental mortality is associated with increasing the optimum investment in maintenance, reducing fecundity, and increasing lifespan. The first, second and fourth columns of table 1 reveal correlations of the type that are familiar from the study of natural populations. What the disposable soma theory adds is the third column. This is an explicit prediction of the mechanism by which these correlations are generated physiologically, namely by varying the optimal investment in maintenance.

The level of environmental mortality plays a similarly central role in the evolution of lifespan in terms of antagonistic pleiotropy, although this theory's predictions with regard to the mechanism of lifespan determination are less explicit. Because most deaths in iteroparous adults are due to environmental causes, as already noted, the dominant factor influencing the rate of decline in the force of natural selection is the level of adult environmental mortality. The rate of decline in the force of natural selection is the key to determining what age is meant by 'late', in relation to the late deleterious fitness effects of antagonistically pleiotropic genes. Therefore, a high level of environmental mortality should be associated with short lifespan, and vice versa.

4. COSTS OF MAINTENANCE AND REPRODUCTION

The idea of costs of maintenance and reproduction has been used in this paper until now without specifying the sources and possible magnitudes of these costs. In the disposable soma theory, these costs form the basis of the theory and therefore play an essential part. In the antagonistic pleiotropy theory, as originally described by Williams (1957), a more general concept of pleiotropy was advanced. As an example, Williams cited a mutation arising that has a favourable effect on the calcification of bone in the developmental period but which expresses itself in a subsequent somatic environment in the calcification of the connective tissue of the arteries. During recent years, however, advocacy of the antagonistic pleiotropy theory has emphasized the trade-offs between survival and fecundity. With this shift in focus the antagonistic pleiotropy theory has become rather similar to the disposable soma theory and now also rests heavily on the idea of costs of maintenance and reproduction.

The recognition of a cost of reproduction has long been a major theme in ecology (Stearns 1976; Bell 1980; Partridge & Harvey 1988). Patently, there are substantial costs in the production of progeny and in the repertoire of morphological adaptations and behaviours associated with reproduction.

Costs of maintenance, although mentioned passingly as the obverse of reproductive effort, have received considerably less attention. Kirkwood (1981) reviewed the general problem of the evolution of maintenance and repair capabilities. For any repair or maintenance process to evolve, three basic conditions must be fulfilled. First, the organism must be able to survive the damage at least for long enough for repair to take place. Secondly, the information to restore the damaged part to its undamaged form must be available (the requirement of 'repeatability'). Thirdly, the overall benefit of repair, in terms of its effect on fitness must outweigh its cost. For severe forms of damage the fulfilment of all three conditions is less likely than for minor damage. Thus, the phylogenic distribution of major repair functions is likely to be less uniform than for correction of minor defects. This is presumably the reason why regeneration ability varies markedly between species (Kirkwood 1981; Reichman 1984). Basic maintenance processes, on the other hand, are expected to be more evenly distributed, and we observe similar mechanisms in virtually all species.

Maintenance costs arise at three principal levels. The first comprises the costs incurred in those aspects of the construction of non-renewing parts, like teeth, which are concerned with durability. It might be argued that these are not really costs of maintenance, as such. They are, however, a part of the overall maintenance picture so we include them here. The second level comprises the cost of maintenance involving cell renewal. The way skin maintains itself is one example, the immune system another. The third level involves the processes of intracellular maintenance. These are particularly important because of their ubiquity. All cells require them, and there are striking similarities between basic cell maintenance processes among very different forms of life. Basic cell maintenance processes are also being found to be quite costly (see Kirkwood *et al.* 1986). Maintenance of DNA, for instance, involves several different repair systems, each tailored for particular types of damage (Sedgwick 1986). The fidelity of DNA replication is maintained by the highly tuned proofreading capacity of DNA polymerases. Proofreading is metabolically expensive, as can be shown by *in vitro* studies on mutant polymerases with altered proofreading capabilities (see Kirkwood *et al.* 1986, pp. 5–6). Accuracy in protein synthesis is also important in maintaining cell viability, and this too is costly. The cost of proofreading the charging of tRNA by aminoacyl-tRNA synthetases has been estimated to account for as much as 2% of the energy budget of a cell (Savageau & Freter 1979), and this is just one of several operations in accurate protein synthesis. Similarly, there are energy-consuming enzymic processes to degrade abnormal proteins (Hipkiss 1989) and to protect cells against the highly reactive oxygen radicals that arise as by-products of aerobic metabolism (Halliwell & Gutteridge 1989).

Given that each of these maintenance processes has its costs, some quite considerable, we expect the investment in them to be optimized at a level below that which permits indefinite survival. What we also expect is some degree of harmonization between rates of accumulation of different types of damage, which is probably part of the reason why attempts to explain senescence in terms of just DNA damage, or just protein errors, for example, have yielded inconclusive results (see, for example, Warner *et al.* 1987).

5. TESTS OF THE EVOLUTIONARY THEORIES

(*a*) *Evolutionary versus non-evolutionary theories*

Non-evolutionary theories of senescence are those which suggest that senescence does not require an evolutionary explanation and arises, for example, as the inevitable result of wear-and-tear. Although there are logical arguments that can be advanced against the inevitability of wear-and-tear in a biological organism (see, for example, Williams 1957), the evolutionary theories could be falsified if it was shown that the presence and absence of senescence failed to correlate with the basic requirements of the evolutionary theories.

Bell (1984) addressed this point. By examining several freshwater invertebrates, Bell found that those species that reproduced vegetatively did not senesce, whereas those that laid eggs did. This confirms the importance of the soma/germ-line distinction and supports the idea that senescence evolves as a consequence of natural selection acting upon age-specific rates of reproduction and survival.

(*b*) *Comparative studies*

A second approach to testing the evolutionary theories is through comparative studies, particularly of somatic maintenance mechanisms. In an often-cited study Hart & Setlow (1974) found a strong positive correlation between a particular form of DNA repair – excision repair of pyrimidine dimers – and lifespan, and this result has been confirmed in other laboratories (see Rattan 1989). However, the criticism can be made of these and most other comparative studies to date that they have failed to take proper account of potentially confounding variables, such as body size, age of cell donor, and biopsy site.

In spite of the limitations of current comparative data on somatic maintenance mechanisms, it is probably safe to conclude that long-lived organisms look after their cells better than short-lived organisms, one clear example being that the lifetime risk of cancer is similar for mice and humans, despite their great differences in body size and lifespan. However, better comparative studies would be welcome.

(*c*) *Selection for altered lifespan*

The third approach to testing the evolutionary theories is to select for altered lifespan. Because many things can shorten life, and not all of these have to do with senescence, the most interesting challenge is to select for increased lifespan. This was first done successfully using *Drosophila* spp. by Wattiaux (1968*a*, *b*), and has since been repeated several times in *Drosophila melanogaster* (e.g. Rose & Charlesworth 1980, 1981*b*; Rose 1984; Luckinbill & Clare 1985). The basic procedure followed in these experiments has been to select indirectly for longevity by selecting for late female fecundity. This is particularly likely to work where there is a trade-off between early fecundity and late survival, and Rose & Charlesworth (1981*a*) showed that in outbred populations the necessary negative genetic covariance in life-history characters appears to exist. The importance of using outbred populations is both to ensure the necessary genetic variation and to avoid inbreeding depression. Inbreeding depression can cause artifactual strong positive correlations between survival and fecundity (see Giesel 1979; Rose 1984*a*).

The general experimental strategy involves raising two sets of populations. The 'young' control lines are reproduced under standard conditions, breeding from young females. The 'old' lines are reproduced from old females by selecting only eggs laid towards the end of the reproductive period. Because the selection does in fact work, the egg collection in the 'old' lines can take place later and later in successive generations.

A necessary feature of the experimental procedure is that the culture must be grown at high larval density (Luckinbill & Clare 1985; Clare & Luckinbill 1985). Failure to control the larval density among 'young' and 'old' lines can produce genotype-by-environment interactions that interfere with the selection for postponed senescence. These probably explain earlier discordant results of Lints & Hoste (1974, 1977). The need to control larval density at high rather than low levels is that when exposed to a rich, non-competitive environment, the larvae respond with rapid growth and early reproduction, vitiating the effects of selection for late fecundity. Replication of the selected lines is also important because if there are only one or two selected lines used for comparison, then they may have spuriously differentiated in response to selection due to linkage and related finite-population effects.

A study by Rose (1984*b*) gave results typical of the above selection procedure. Rose (1984*b*) found that (i) mean adult lifespan in the 'old' lines was increased compared with the 'young' lines, and (ii) females in the 'old' lines laid fewer eggs early in life and somewhat more eggs late in life than the females in the 'young' lines. The mean lifespan for females in the 'old' lines of Rose (1984*b*) was increased by 29% after 15 generations of selection. A smaller increase was seen in the mean lifespan of males.

Study of morphological differences between 'young' and 'old' lines revealed no overall size change (Rose *et al.* 1984; Luckinbill *et al.* 1988*a*). Early ovary mass in the 'old' lines was found to be substantially reduced (Rose *et al.* 1984), consistent with the reduced early fecundity. Studies of stress resistance in the 'old' lines (Service *et al.* 1985; Service 1987) found that postponed senescence was associated with increased resistance to starvation, desiccation, and low levels of ambient alcohol. In addition, flight duration is increased in the longer-lived lines (Luckinbill *et al.* 1988*a*; Graves *et al.* 1988; Graves & Rose 1990).

The findings from the selection experiments in *Drosophila* provide a body of evidence consistent with the idea that selection has exploited a trade-off between survival and fecundity. This fits with the predictions of the antagonistic pleiotropy theory, and will also be consistent with the disposable soma theory if it turns out that the reason for the increased longevity in the old lines is that the flies invest more resources in somatic maintenance.

6. LEVELS OF TRADE-OFF BETWEEN SURVIVAL AND FECUNDITY

Theory and evidence support the idea of trade-offs between late survival and early fecundity. But at what level might these trade-offs be operating?

First, the trade-offs might be non-metabolic. For instance, if reproductive activity reduces survival by increased risk exposure, or wear-and-tear, then simply rescheduling fecundity to later ages will increase survivorship. Secondly, the trade-offs might be direct metabolic trade-offs. These are trade-offs that can happen if reproduction and maintenance draw directly from the same supply of resources within the organism, so that reducing the demand for one automatically increases the supply to the other. Thirdly, the trade-offs might be indirect metabolic trade-offs, such that resources are shared but not directly convertible between reproduction and maintenance. For instance, reducing the level of proofreading DNA replication may mean that cells consume less energy in maintenance, but it does not necessarily mean that the organism can produce larger or more frequent litters. Each investment is independently regulated, and reducing one merely provides a more favourable opportunity for adaptations increasing the other.

In general, non-metabolic and direct metabolic trade-offs are likely to be associated with greater plasticity of the life history, both genotypic and phenotypic. Indirect metabolic trade-offs will be slower to respond to altered circumstances, including those imposed in selection experiments.

With this perspective, what might be going on in the selection for postponed senescence in *Drosophila melanogaster*? It has been known for some time that in *Drosophila* spp. sexual activity and reproduction are directly detrimental to survival (Maynard Smith 1958; Kidwell & Malick 1967; Partridge & Farquhar 1981; Partridge *et al.* 1986, 1987; Fowler & Partridge 1989). This non-metabolic trade-off, whatever its physiological basis, means that simply postponing fecundity in the 'old' lines may be a major contributor to their increased mean lifespan. Luckinbill *et al.* (1987) reported that a single genetic factor appeared to be mainly responsible for the delay in senescence, consistent with a predominating non-metabolic trade-off. This estimate rested, however, on a method of calculation which questionably assumed an equal influence of the genes measured. Luckinbill *et al.* (1988*b*) later performed a more direct study using chromosome substitution, which suggested that longevity is under polygenic control with contributing elements on all chromosomes. However, one chromosome, the third chromosome, was found to be by far the most influential, accounting for two thirds of the observed effect in females.

The exact basis of the trade-offs between survival and fecundity in *Drosophila* remains to be discovered, and as Luckinbill *et al.* (1988*b*) point out, this will depend on locating and functionally analysing the individual genes that are involved. We note here, however, that although selection for postponed senescence has proved valuable in generating lines to compare, selection will expose most strongly those trade-offs that are most amenable to change and which respond quickest. These will tell us part, but not necessarily all of the story.

7. DISCUSSION

Two different approaches to understanding evolution of senescence lead to the same general conclusion, namely that a major factor is likely to have been the sacrifice of late survival in favour of enhanced early reproduction. One approach is through population genetics, particularly inspired by Medawar (1952) and

Williams (1957). The foundation of this approach is the recognition that the force of selection declines with age. Because of this, genes that have beneficial effects early in life can be favoured by selection even if they produce major deleterious effects later on (antagonistic pleiotropy). The second approach stems from studies on the physiology of senescence, particularly of theories proposing that stochastic molecular damage is responsible. From asking how much an organism ought to invest in maintaining itself comes the disposable soma theory (Kirkwood 1977, 1981; Kirkwood & Holliday 1979), which asserts that the optimum investment in maintenance is less than the mimimum level required for indefinite survival.

The difference between the disposable soma theory and antagonistic pleiotropy is partly a difference between an optimality theory approach (see Parker & Maynard Smith 1990) and a quantitative genetics approach, and partly about the level at which the two theories seek to explain senescence. The disposable soma theory assumes a trade-off of resources between maintenance (survival) and reproduction and shows that with this assumption the evolutionary optimum leads directly to senescence. The theory gives broad insights into the physiological basis of senescence and the genetic control of longevity, and it thus combines ultimate and proximate factors in a unified theory. The antagonistic pleiotropy theory assumes a more formal principle that trade-offs can occur between early beneficial and late deleterious effects of genes influencing the life history; the nature of the genes' action is not specified.

Is the disposable soma theory a causal subset of the antagonistic pleiotropy theory? One can answer yes to this question if the genes responsible for somatic maintenance functions are regarded as having the pleiotropic effects that they (i) prolong survival and (ii) consume resources which might otherwise be used for reproduction. It is then necessary to invert these effects to fit the antagonistic pleiotropy theory. In other words, it is by depressing the action of somatic maintenance genes that the benefit of enhanced early reproduction is generated at the expense of late senescence. The optimality approach obviates the need for such indirect reasoning. The two theories are more accurately seen as complementary than overlapping. The antagonistic pleiotropy theory is appropriate to the study of variance and covariance of life-history characters within a population and the effects of selection upon them. The disposable soma theory leads more directly to predictions about physiological and comparative aspects of longevity and senescence.

In suggesting that the sacrifice of late survival for early reproduction is the major factor in the evolution of senescence, we must not lose sight of other explanations. For instance, there may be pleiotropic effects not involving the trade-off of survival for fecundity, as was clearly envisaged by Williams (1957). Also, the iteroparous life history always provides scope for mutation accumulation, even if this is not the main driving force in the evolution of senescence. Mueller (1987) reported that preventing *Drosophila melanogaster* from breeding late led to reduced late fecundity, which he suggested was best explained as the result of mutation accumulation, arising from the absence of selection against mutations affecting late fecundity. Service *et al.* (1988) have also reported results in *D. melanogaster* consistent with mutation accumulation. After selecting successfully for postponed senescence, they reversed the direction of selection in some old lines. The original survival and fecundity patterns were restored, but some of the increased stress resistance remained unaltered. This, they suggested, was because the original selection had shifted some late-acting deleterious effects which were the result of mutation accumulation rather than pleiotropy.

The success of selection for postponed senescence in *Drosophila* has confirmed one of the main predictions of the theory that senescence has evolved by trading late survival for early reproduction. It has also provided us with a valuable experimental model to study senescence in this species. Just how far this model will take us remains to be seen. Recent studies by K. Fowler & L. Partridge (unpublished data) have reported a different pattern of response to selection for postponed senescence, where 'old' and 'young' did not differ in fecundity early in the lifespan, and where it was also found that the larval period was affected, with increased larval development time and lowered larval survival under competitive conditions. The effects of selection and the nature of the trade-offs that result need further study, including the effects on the larval stage of the life cycle.

Can similar selection work in other species, such as the mouse? Theory predicts it should, but it will of course be a much longer and more expensive experiment. Should we attempt it? Serious discussion of the feasibility of such a project has been taking place (Charlesworth 1988; Johnson 1988; Rose 1988, 1990). Will the response per generation of selection be as fast as in *Drosophila*? This will depend on the genetic variance and covariance in the population, and on the levels of trade-off which can be exploited.

Although selection experiments provide one avenue to explore the genetic basis of senescence, other avenues are provided by careful comparative studies between different species (see, for example, Sacher & Hart 1978), by genetic analysis of longevity mutants within a single species (see, for example, Johnson 1987), and by gene transfer studies in transgenic animals (see, for example, Epstein *et al.* 1987). The evolutionary view described in this paper leads us to predict that genes involved in regulating the trade-offs between costs of maintenance and costs of reproduction should be principal candidates for intensive study. The disposable soma theory leads us to understand senescence as the result of tuning the investment in somatic maintenance at a level that is enough to survive the natural expectation of life in the wild, but not higher. This is a prediction which is eminently testable. We must recognize, however, that evolutionary theory also tells us that in iteroparous organisms no single physiological process is likely to cause senescence on its own. The practical problem of teasing out individual contributions to the overall process of senescence remains a major challenge.

APPENDIX

This appendix outlines the model used to obtain the results in figure 1 and table 1 (this is a modified version of an earlier model by Kirkwood & Holliday (1986) and Kirkwood (1990)). Survivorship $l(x;s)$ and fecundity $m(x;s)$ are defined as functions that depend on age x and investment in maintenance s. The form of $l(x;s)$ is determined by specifying juvenile and adult mortality. The age-distribution of juvenile mortality does not matter in the model and total juvenile mortality is defined as V, independent of s. Adult mortality rate μ is assumed to follow the Gompertz–Makeham equation

$$\mu = \alpha\, e^{\beta x} + \gamma,$$

where the exponential term represents intrinsic, age-dependent mortality, and the constant γ represents extrinsic, age-independent 'environmental mortality'. When $\beta > 0$, the adult mortality rate increases with age and senescence occurs. This mortality pattern provides a good fit to survival data in several species (Sacher 1978; Finch *et al.* 1990). The resulting form for $l(x;s)$ is

$$l(x;s) = (1-V)\exp\left[-\alpha(e^{\beta x}-e^{\beta a})/\beta-\gamma(x-a)\right].$$

The dependence on s is introduced by making β a decreasing function that reaches zero at $s = s'$ (e.g. $\beta = \beta_0(s'/s-1)$ for $s < s'$; $\beta = 0$ for $s \geqslant s'$). This has the required property that senescence occurs more slowly as s is increased, until for $s \geqslant s'$ senescence does not occur at all.

Fecundity is specified by assuming that reproduction begins at peak rate h at age a, and that if $\beta > 0$ (i.e. if senescence occurs) the reproductive rate declines like a survival curve driven by the intrinsic, exponential component of the adult mortality rate. This has the advantage of linking the effects of senescence on survival and fecundity through the same parameter β, and gives

$$m(x;s) = h\exp\left[-\alpha(e^{\beta x}-e^{\beta a})/\beta\right] \quad \text{for} \quad x \geqslant a.$$

The parameters h and a are made to depend on s, because increasing the investment in maintenance leaves fewer resources for growth and reproduction, and vice versa. For example, we assume $a = a_{min}/(1-s)$ and $h = h_{max}(1-s)$. Provided that β and h are decreasing functions of s, and a is an increasing function of s, the precise forms of these functions are not critical for the general pattern of results which is obtained.

The example of the mouse (table 1, centre row) was fitted using data on age at first reproduction a (taken as 6 weeks), maximum reproductive rate h (taken as 1.0 birth female^{-1} week^{-1}), and the pattern of intrinsic age-dependent adult mortality (taken as a Gompertz function of the form $0.01e^{0.1x}$, where x is age in months; see Sacher (1978)). From these data, and applying the constraint $r = 0$ for $s = s^*$, values for juvenile mortality (95%), environmental mortality ($\gamma = 0.032$), and lifespan (36 months) were determined. The value of s' was set at 0.8, and the value of s^* found to be 0.5. A degree of freedom is available to define the arbitrary scale for s so that s^* takes a convenient reference value, but whatever scaling is chosen s^* is always well below s'.

REFERENCES

Bell, G. 1980 The costs of reproduction and their consequences. *Am. Nat.* **116**, 45–76.

Bell, G. 1984 Evolutionary and nonevolutionary theories of senescence. *Am. Nat.* **124**, 600–603.

Charlesworth, B. 1980 *Evolution in age-structured populations.* Cambridge University Press.

Charlesworth, B. 1988 Selection for longer-lived rodents. *Growth, Dev. Aging* **52**, 211.

Clare, M. J. & Luckinbill, L. S. 1985 The effects of gene–environment interaction on the expression of longevity. *Heredity, Lond.* **55**, 19–29.

Cole, L. C. 1954 The population consequences of life history phenomena. *Q. Rev. Biol.* **29**, 103–137.

Comfort, A. 1979 *The biology of senescence*, 3rd edn. Edinburgh: Churchill Livingstone.

Epstein, C. J., Avraham, K. B., Lovett, M., Smith, S., Elroy-Stein, O., Rotman, G., Bry, C. & Groner, G. 1987 Transgenic mice with increased Cu/Zn-superoxide dismutase activity: animal model of dosage effects in Down syndrome. *Proc. natn. Acad. Sci. U.S.A.* **84**, 8044–8048.

Ewens, W. J. 1979 *Mathematical population genetics.* Berlin: Springer-Verlag.

Finch, C. E. & Schneider, E. L. 1985 *Handbook of the biology of aging.* 2nd edn. New York: Van Nostrand Rheinhold.

Finch, C. E., Pike, M. C. & Witten, M. 1990 Slow mortality rate accelerations during aging in some animals approximate that of humans. *Science, Wash.* **249**, 902–905.

Fisher, R. A. 1930 *The genetical theory of natural selection.* Oxford University Press. (Page numbers in text refer to the 1958 2nd edn, New York: Dover.)

Fowler, K. & Partridge, L. 1989 A cost of mating in female fruitflies. *Nature, Lond.* **338**, 760–761.

Giesel, J. T. 1979 Genetic co-variation of survivorship and other fitness indices in *Drosophila melanogaster. Exp. Geront.* **14**, 323–328.

Graves, J. L. & Rose, M. R. 1990 Flight duration in *Drosophila melanogaster* selected for postponed senescence. In *Genetic effects on aging* II (ed. D. E. Harrison), pp. 59–65. Caldwell, New Jersey: Telford Press.

Graves, J. L., Luckinbill, L. S. & Nichols, A. 1988 Flight duration and wing beat frequency in long- and short-lived *Drosophila melanogaster. J. Insect Physiol.* **34**, 1021–1026.

Haldane, J. B. S. 1941 *New paths in genetics.* London: George Allen & Unwin.

Halliwell, B. & Gutteridge, J. M. C. 1989 *Free radicals in biology and medicine*, 2nd edn. Oxford University Press.

Hamilton, W. D. 1966 The moulding of senescence by natural selection. *J. theor. Biol.* **12**, 12–45.

Hart, R. W. & Setlow, R. B. 1974 Correlation between deoxyribonucleic acid excision-repair and lifespan in a number of mammalian species. *Proc. natn. Acad. Sci. U.S.A.* **71**, 2169–2173.

Hipkiss, A. R. 1989 The production and removal of abnormal proteins: a key question in the biology of ageing. In *Human ageing and later life* (ed. A. M. Warnes), pp. 15–28. London: Edward Arnold.

Johnson, T. E. 1987 Aging can be genetically dissected into component processes using long-lived lines of *Caenorhabditis elegans. Proc. natn. Acad. Sci. U.S.A.* **84**, 3777–3781.

Johnson, T. E. 1988 Thoughts on the selection of longer-lived rodents. *Growth, Dev. Aging* **52**, 207–209.

Kidwell, J. F. & Malick, L. E. 1967 The effect of genotype, mating status, weight and egg production on longevity in *Drosophila melanogaster. J. Hered.* **58**, 169–172.

Kirkwood, T. B. L. 1977 Evolution of ageing. *Nature, Lond.* **270**, 301–304.
Kirkwood, T. B. L. 1981 Repair and its evolution: survival versus reproduction. In *Physiological ecology: an evolutionary approach to resource use* (ed. C. R. Townsend & P. Calow), pp. 165–189. Oxford: Blackwell Scientific Publications.
Kirkwood, T. B. L. 1985 Comparative and evolutionary aspects of longevity. In *Handbook of the biology of aging*, 2nd edn (ed. C. E. Finch & E. L. Schneider), pp. 27–44. New York: Van Nostrand Reinhold.
Kirkwood, T. B. L. 1990 The disposable soma theory of aging. In *Genetic effects on aging* II (ed. D. E. Harrison), pp. 9–19. Caldwell, New Jersey: Telford Press.
Kirkwood, T. B. L. & Cremer, T. 1982 Cytogerontology since 1881: a reappraisal of August Weismann and a review of modern progress. *Hum. Genet.* **60**, 101–121.
Kirkwood, T. B. L. & Holliday, R. 1979 The evolution of ageing and longevity. *Proc. R. Soc. Lond.* B **205**, 531–546.
Kirkwood, T. B. L. & Holliday, R. 1986 Ageing as a consequence of natural selection. In *The biology of human ageing* (ed. A. J. Collins & A. H. Bittles), pp. 1–16. Cambridge University Press.
Kirkwood, T. B. L., Rosenberger, R. F. & Galas, D. J. 1986 *Accuracy in molecular processes: its control and relevance to living systems*. London: Chapman & Hall.
Lack, D. 1954 *The natural regulation of animal numbers*. Oxford University Press.
Lints, F. A. & Hoste, C. 1974 The Lansing effect revisited. I. Life-span. *Exp. Geront.* **9**, 51–69.
Lints, F. A. & Hoste, C. 1977 The Lansing effect revisited. II. Cumulative and spontaneously reversible parental age effects on fecundity in *Drosophila melanogaster*. *Evolution* **31**, 387–404.
Luckinbill, L. S. & Clare, M. J. 1985 Selection for life span in *Drosophila melanogaster*. *Heredity, Lond.* **55**, 9–18.
Luckinbill, L. S., Clare, M. J., Krell, W. L., Cirocco, W. C. & Richards, P. A. 1987 Estimating the number of genetic elements that defer senescence in *Drosophila*. *Evol. Ecol.* **1**, 37–46.
Luckinbill, L. S., Graves, J. L., Tomkiw, A. & Sowirka, O. 1988*a* A qualitative analysis of some life-history correlates of longevity in *Drosophila melanogaster*. *Evol. Ecol.* **2**, 85–94.
Luckinbill, L. S., Graves, J. L., Reed, A. H. & Koetsawang, S. 1988*b* Localizing genes that defer senescence in *Drosophila melanogaster*. *Heredity, Lond.* **60**, 367–374.
Maynard Smith, J. 1958 The effects of temperature and of egg-laying on the longevity of *Drosophila subobscura*. *J. exp. Biol.* **35**, 832–842.
Maynard Smith, J. 1962 Review lectures on senescence. I. The causes of ageing. *Proc. R. Soc. Lond.* B **157**, 115–127.
Medawar, P. B. 1946 Old age and natural death. *Mod. Quart.* **2** (new series), 30–49.
Medawar, P. B. 1952 *An unsolved problem of biology*. London: Lewis.
Medawar, P. B. 1955 The definition and measurement of senescence. In *Ciba Foundation Colloquia on ageing*, vol. 1 (ed. G. E. W. Wolstenholme & M. P. Cameron), pp. 4–15. London: Churchill.
Medvedev, Z. A. 1990 An attempt at a rational classification of theories of ageing. *Biol. Rev.* **65**, 375–398.
Mueller, L. D. 1987 Evolution of accelerated senescence in laboratory populations of *Drosophila*. *Proc. natn. Acad. Sci. U.S.A.* **84**, 1974–1977.
Parker, G. A. & Maynard Smith, J. 1990 Optimality theory in evolutionary biology. *Nature, Lond.* **348**, 27–33.
Partridge, L. & Farquhar, M. 1981 Sexual activity reduces longevity of male fruitflies. *Nature, Lond.* **294**, 580–582.
Partridge, L. & Harvey, P. H. 1988 The ecological context of life-history evolution. *Science, Wash.* **241**, 1449–1455.
Partridge, L., Fowler, K., Trevitt, S. & Sharp, W. 1986 An examination of the effects of males on the survival and egg-production rates of female *Drosophila melanogaster*. *J. Insect Physiol.* **33**, 745–749.
Promislow, D. 1991 The evolution of senescence in mammals: comparative evidence. *Evolution*. (submitted.)
Rattan, S. I. S. 1989 DNA damage and repair during cellular ageing. *Int. Rev. Cytol.* **116**, 47–88.
Reichman, O. J. 1984 Evolution of regeneration capabilities. *Am. Nat.* **123**, 752–763.
Robertson, O. H. 1961 Prolongation of the lifespan of Kokanee salmon (*Oncorhynkus nerka kennerlyi*) by castration before beginning of gonad development. *Proc. natn. Acad. Sci. U.S.A.* **47**, 609–621.
Rose, M. R. 1982 Antagonistic pleiotropy, dominance, and genetic variation. *Heredity, Lond.* **48**, 63–78.
Rose, M. R. 1984*a* Genetic covariation in *Drosophila* life history: untangling the data. *Am. Nat.* **123**, 565–569.
Rose, M. R. 1984*b* Laboratory evolution of postponed senescence in *Drosophila melanogaster*. *Evolution* **38**, 1004–1010.
Rose, M. R. 1985 Life history evolution with antagonistic pleiotropy and overlapping generations. *Theor. Popul. Biol.* **28**, 342–358.
Rose, M. R. 1988 Response to 'Thoughts on the selection of longer-lived rodents' *Growth, Dev. Aging* **52**, 209–211.
Rose, M. R. 1990 Should mice be selected for postponed aging? A workshop summary. *Growth, Dev. Aging* **54**, 7–15.
Rose, M. R. 1991 *Evolutionary biology of aging*. New York: Oxford University Press.
Rose, M. R. & Charlesworth, B. 1980 A test of evolutionary theories of senescence. *Nature Lond.* **287**, 141–142.
Rose, M. R. & Charlesworth, B. 1981*a* Genetics of life history in *Drosophila melanogaster*. I. Sib analysis of adult females. *Genetics* **97**, 173–186.
Rose, M. R. & Charlesworth, B. 1981*b* Genetics of life history in *Drosophila melanogaster*. II. Exploratory selection experiments. *Genetics* **97**, 173–186.
Rose, M. R., Dorey, M. L., Coyle, A. M. & Service, P. M. 1984 The morphology of postponed senescence in *Drosophila melanogaster*. *Can. J. Zool.* **62**, 1576–1580.
Sacher, G. A. 1978 Evolution of longevity and survival characteristics in mammals. In *The genetics of aging* (ed. E. L. Schneider), pp. 151–167. New York: Plenum.
Sacher, G. A. & Hart, R. W. 1978 Longevity, aging and comparative cellular and molecular biology of the house mouse, *Mus musculus*, and the white-footed mouse, *Peromyscus leucopus*. In *Genetic effects on aging* (ed. D. Bergsma & D. E. Harrison), pp. 71–96. New York: Alan Liss.
Savageau, M. A. & Freter, R. R. 1979 On the evolution of accuracy and cost of proofreading tRNA aminoacylation. *Proc. natn. Acad. Sci. U.S.A.* **76**, 4507–4510.
Sedgwick, S. G. 1986 Stability and change through DNA repair. In *Accuracy in molecular processes: its control and relevance to living systems* (ed. T. B. L. Kirkwood, R. F. Rosenberger & D. J. Galas), pp. 233–289. London: Chapman & Hall.
Service, P. M. 1987 Physiological mechanisms of increased stress resistance in *Drosophila melanogaster* selected for postponed senescence. *Physiol. Zool.* **60**, 321–326.
Service, P. M., Hutchinson. E. W., MacKinley, M. D. & Rose, M. R. 1985 Resistance to environmental stress in *Drosophila melanogaster* selected for postponed senescence. *Physiol. Zool.* **58**, 380–389.
Service, P. M., Hutchinson. E. W. & Rose, M. R. 1988 Multiple genetic mechanisms for the evolution of senescence in *Drosophila melanogaster*. *Evolution* **42**, 708–716.
Sibly, R. M. & Calow, P. 1986 *Physiological ecology of*

animals: an evolutionary approach. Oxford: Blackwell Scientific Publications.
Stearns, S. C. 1976 Life-history tactics: a review of the ideas. *Q. Rev. Biol.* **51**, 3–47.
Townsend, C. R. & Calow, P. 1981 *Physiological ecology: an evolutionary approach to resource use.* Oxford: Blackwell Scientific Publications.
Warner, H. R., Butler, R. N., Sprott, R. L. & Schneider, E. L. 1987 *Modern biological theories of aging.* New York: Raven Press.
Wattiaux, J. M. 1968*a* Parental age effects in *D. pseudoobscura. Exp. Gerontol.* **3**, 55–61.
Wattiaux, J. M. 1968*b* Cumulative parental age effects in *D. subobscura. Evolution* **22**, 406–421.
Williams, G. C. 1957 Pleiotropy, natural selection and the evolution of senescence. *Evolution* **11**, 398–411.
Wodinsky, J. 1977 Hormonal inhibition of feeding and death in *Octopus*: control by optic gland secretion. *Science, Wash.* **198**, 948–951.

Ecological determinants of life-history evolution

LAURENCE D. MUELLER

Department of Ecology and Evolutionary Biology, University of California, Irvine, California 92717, U.S.A.

SUMMARY

Density-dependent natural selection has been studied, empirically with laboratory populations of *Drosophila melanogaster*. Populations kept at very high and low population density have become differentiated with respect to important fitness-related traits. There is now some understanding of the behavioural and physiological basis of these differences. These studies have identified larval competitive ability and efficiency of food utilization as traits that are negatively correlated with respect to effects on fitness. Theory that illuminates and motivates additional research with this experimental system has been lacking. Current research has focused on models that incorporate many details of *Drosophila* ecology in laboratory environments.

1. INTRODUCTION

The importance of the natural environment in determining the evolution of species was obvious to Darwin. Despite this, the early synthesis of evolutionary biology largely ignored the interaction between ecology and evolution. The first important theoretical fusion of ecological theory and evolutionary theory came with MacArthur (1962) and MacArthur & Wilson (1967).

MacArthur & Wilson utilized the well developed theory of density-dependent population growth to study the outcome of adaptation to extreme population densities. This was a logical first step since population density is a predictable component of the environment and thus amenable to this sort of theoretical investigation. MacArthur & Wilson selected one general model of population growth, the logistic equation, to develop their theory. The great generality of the logistic was also a liability for the purposes of designing tests of the theory.

The two parameters of the logistic, r and K, were variously interpreted as measuring life-history traits that they had little connection to. The result of such a misapplication of theory was the general dismissal of r- and K-selection, as the MacArthur–Wilson theory is often called. In fact, it has become clear that population density can have a profound effect on aspects of life-history evolution. It is also evident that precise theoretical predictions about life-history evolution for any particular species will have to utilize theory that takes into account relevant ecological details of that species.

In the next section I describe a series of studies with laboratory populations of *Drosophila melanogaster* designed to investigate the effects of adaptation to extreme densities. This will be followed by a discussion of the interaction between theory and experiment in this field.

2. EXPERIMENTAL RESEARCH

The early 1970s saw numerous attempts to test the theory of density-dependent natural selection or r- and K-selection as it was sometimes called. These studies used natural populations of different species (McNaughton 1975), or different populations of the same species (Gadgil & Solbrig 1972), which were thought to have experienced different degrees of density regulation. A significant shortcoming of this type of research is that the study populations might differ with respect to environmental variables other than density in an unknown and uncontrolled fashion. Often the inferred differences in the density régimes were uncertain.

(*a*) *Experimental* Drosophila *populations*

To remedy these problems I undertook, in 1978, a study of laboratory populations of *Drosophila melanogaster*. With this organism, replicate samples could be taken from the same source population, thus insuring that initially the populations would be identical with respect to the studied characters. The replicate populations could then be subjected to different density regimes while keeping all other variables constant.

The creation of these populations and important historical events during their maintenance are shown in figure 1. Three replicates of the low density populations (r-populations) were created and maintained using a reproductive population of 50 adults. The high density, K-populations, were maintained at adult population sizes of approximately 1000. In the r-populations both adults and larvae were uncrowded whereas larval and adult crowding were severe in the K-populations. In addition to these density differences were differences in the timing of adult reproduction. In the r-populations adults were generally 3–6 days old at the time of egg laying. Adults in the K-populations

Phil. Trans. R. Soc. Lond. B (1991) **332**, 25–30
Printed in Great Britain

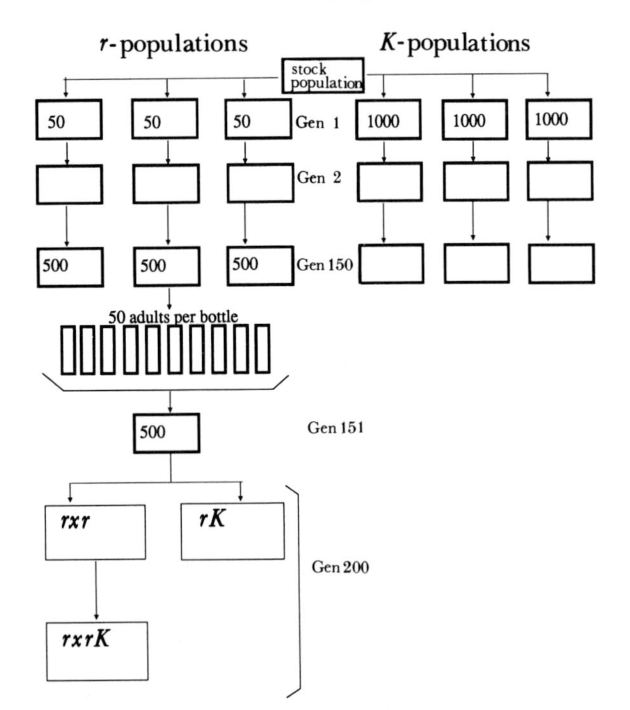

Figure 1. The derivation and maintenance of the *r*-, *K*-, *rxr*-, *rK*- and *rxrK*-populations.

were allowed to breed indefinitely, although the average lifespan was probably only 2–3 weeks. Although differences in adult age at reproduction are unlikely to affect the evolution of larval characters they are important for adult characters. Evidence that age-specific female fecundity has differentiated as a consequence of the timing of adult reproduction in the *r*- and *K*-populations is described in Mueller (1987). After approximately 150 generations the maintenance régime of the *r*-populations was altered such that the effective population size per population was closer to 500 rather than 50; future effects of random genetic drift could therefore be forestalled.

At about generation, 200, three new types of populations were created (figure 1). The *rK*-populations were derived from the three replicate *r*-populations. The *rK*-populations were kept in the *K*-environments and thus these populations represented a form of reverse selection. The *rxr*-populations were created by making all possible pairwise crosses between the three *r*-populations and taking equal numbers of progeny from each cross to initiate each *rxr*-population. These populations were maintained in the same way as the *r*-populations but had genetic variation from all the original populations combined into each constituent population, thus mitigating inbreeding effects that might have been present in the *r*-populations. Finally, from the *rxr*-populations three *rxrK*-populations were derived which were kept in a *K*-population maintenance regime (figure 1).

(*b*) *Phenotypic differentiation of the experimental* Drosophila *populations*

The first test of the *r*- and *K*-populations examined density-dependent rates of population growth at one low (10 adults) and two high (750, 1000) densities (Mueller & Ayala 1981) after eight generations of selection. The results were that at high population density the *K*-populations showed rates of population growth that were significantly elevated relative to the *r*-populations. The opposite result was seen at the low density. These results do not allow us to infer the magnitude or even the direction of change in each individual population from its initial condition although, since the initial condition of these stocks is somewhat arbitrary, the most important result is the differentiation of the *r*- and *K*-populations from each other.

The differences in population growth rates of the *r*- and *K*-populations may be the result of differences in viability of adults or larvae, differences in female fecundity or some combination of these traits. After approximately 30 generations of selection age-specific fecundity and survival rates were estimated at low density and egg-to-adult viability was measured at several larval densities (Bierbaum *et al.* 1989). No differences were seen in the adult fitness components but the *K*-larvae had higher survival rates under crowded larval conditions and the adults which emerged from these crowded cultures and were larger than their *r* counterparts. These results suggest that the differences in population growth rates, at least at high densities, are due to increased survival of *K*-larvae and increased fecundity of *K*-females, as larger females lay more eggs (Robertson 1957; Mueller 1987).

One reason the *K*-larvae survive better under crowded conditions is because these larvae are less likely to pupate on the surface of the food (Mueller & Sweet 1986) where mortality is quite high (Joshi & Mueller 1991). The high densities of larvae used in the Bierbaum *et al.* study create several challenges. Certainly, food is limiting so that a larva that is more efficient at processing food should have a higher probability of survival and should be larger than a less efficient larva all other things being equal. These crowded larvae must also contend with high concentrations of waste products that they inevitably eat. It is known that this consumption of waste has effects on viability and development time (Botella *et al.* 1985) and may also have effects on adult size. Below, I describe experiments aimed at addressing the affects of limited food on viability and adult size. Studies of the effects of waste products on the *r*- and *K*-populations are currently in progress.

Larval food levels may be carefully controlled using techniques first described by Bakker (1961). By examining viability of *r*- and *K*-larvae at various food levels in pure populations and in competition with a genetically marked standard population, it is possible to estimate, for each population, the minimum food level necessary for successful pupation (m) and larval competitive ability for food (α), (Mueller 1988*a*). It is worth noting that although competitive ability should be under strong selection in crowded environments (Mueller 1988*b*), it will not lead to improvements in density-dependent viability and so cannot explain the observations of Bierbaum *et al.* However, reductions in the minimum food requirements will lead to increases in viability (Mueller 1988*b*). Competitive ability showed consistent and large differences between the *r*- and *K*-populations (Mueller 1988*a*). The average competitive ability of the *K*-larvae was 1.14 and for the *r*-larvae 0.72. However, the minimum food requirements showed no consistent differences between the populations.

The estimates of minimum food requirements obtained in the preceding experiments were rather imprecise because this value was inferred from observations made on groups (100) of larvae. These results were checked by a second independent experiment. In this experiment a single larva was placed in a vial with measured quantities of food. With sufficient replication of various food levels this experimental design provides a direct estimate of the minimum amount of food necessary for pupation. This experiment showed that the *K*-populations consistently required more food to pupate successfully than did the *r*-populations (Mueller 1990). These results may show an antagonistic relation between competitive ability and efficiency. Several lines of evidence suggest such a relation.

The feeding rates (measured by rate of cephalopharyngeal contractions) of third instar *K*-larvae are significantly greater than *r*-larvae (Joshi & Mueller 1989). Previous research on this larval behaviour has shown that larvae artificially selected for high feeding rates are better competitors than slow feeders (Burnet *et al.* 1977). Thus it would appear that natural selection for high competitive ability in the *K*-populations has resulted in increasing the feeding rates of larvae. In addition, Sewell *et al.* (1975) have shown that larvae artificially selected for fast feeding are generally more active than slow feeders and that food passes through their alimentary tract more rapidly.

These observations suggest that the fast-feeding *K*-larvae spend more of their energy budget on activity than the slower-feeding *r*-larvae. In addition, the fast rate at which food is processed may prevent larvae from extracting all the available energy (Slansky & Feeny 1977). If these assumptions are correct than a logical consequence of selection for increased competitive ability would be reduced efficiency. Future research should allow us to collect direct evidence on the nature of any physiological differences between the *r*- and *K*-larvae.

(*c*) *Selection versus genetic drift*

One extreme viewpoint of the *r*–*K* differentiation is that in fact density has no effect on life history and the only cause of the differences observed between the populations is because of the small size (during the first 150 generations) of the *r*-populations and the consequent fixation of deleterious mutations. This possibility has been studied by examining the phenotypes of r–F_1 populations. If the inbreeding hypothesis were correct than the r–F_1 offspring should show elevated (near *K*-values) values of the phenotype relative to the depressed values of the parental *r*-populations. For the pupation height, competitive ability and feeding rate phenotypes the r–F_1 progeny showed intermediate values relative to the parental values.

Recently, I have addressed this question from a different perspective. By placing the *r*- and *rxr*-populations in the *K*-environments their response to high densities can be observed. If the *K*-environment places no new selective pressures on the population or if the *r*-populations have been completely depleted of genetic variation then the expectation is that there will be no differences between the *rK*- and *rxrK*-populations and their controls, which are *r* and *rxr* respectively. Pupation height and larval feeding rates appear to have increased in the *rK*- and *rxrK*-populations relative to their controls (P. Z. Guo, L. D. Mueller and F. J. Ayala, unpublished observations). In addition, population growth rates appear to have increased, at high densities, in the *rK*- and *rxrK*-populations relative to controls and decreased at low densities relative to controls. Thus, the initial trade-off in population growth rates noted by Mueller & Ayala (1981) has been confirmed by this reverse selection experiment.

3. THEORY OF DENSITY-DEPENDENT NATURAL SELECTION

(*a*) *Verbal theory*

The ideas of *r*- and *K*-selection developed by MacArthur (1962) and MacArthur & Wilson (1967) contained a mixture of quantitative and verbal theory. This verbal theory was greatly expanded by Pianka (1970, 1972). Although the asset of verbal theory is the ability to pose theory in a simple, intuitive form, the logic behind the verbal theory of *r*- and *K*-selection has often been faulty. One major prediction of the verbal theory of *r*- and *K*-selection is that *K*-selection should favour repeated episodes of breeding or iteroparity. However, the most detailed studies of density-dependent selection in age-structured populations shows that early reproduction will be continually favoured at high population densities (Charlesworth 1980). For instance, assume density dependence results in each female producing only two net progeny. A genotype that has these progeny early in life will, over any fixed time-period, leave more descendants than a genotype that has her two progeny late in life, assuming identical mortality patterns, and thus will become more common even if the total population size is constant.

Another prediction of the verbal theory is that *K*-selection will favour large body size (Pianka 1972). However, density-dependent selection may either lead to increases or decreases in average body size depending on detailed assumptions about the organisms life history (Mueller 1988*b*). An unfortunate consequence of ill-posed verbal theory has been the abandonment of density-dependent selection as an important factor in life-history evolution (Boyce 1984).

(*b*) *Quantitative theories*

(i) *General formulations*

The first models of density-dependent natural selection utilized the logistic model of population growth and assumed that the two parameters of the model, *r* and *K*, were controlled by allelic variation at a single locus (Roughgarden 1971). In practical terms the most useful prediction to come from this model is that density-dependent rates of population growth should respond to natural selection. If there is not a single genotype capable of maximum growth rates at all densities (e.g. a trade-off exists between high r and high K) then the outcome of evolution depends on the environment. This very general prediction has been observed for *D. melanogaster* as described earlier (Mueller & Ayala 1981).

The difficulty with this theory is that its very general nature precludes more detailed predictions concerning the evolution of life-history traits. For instance, even though K should increase in populations kept at high densities, it is difficult to predict precisely which life-histories will be most useful for effecting such an increase without a more detailed specification of the organisms ecology. This does not mean that the general models of density-dependent selection, or any other general model for that matter, are unless, Such models allow one to explore a variety of phenomena and determine whether certain lines of intuition are reasonable. In addition, the components of the model which are important for guiding the process of interest may be identified. However, when tests of such theories are made using a specific organism, it must be kept in mind that this specific creature may have attributes that violate key assumptions of the general model.

The discrete logistic assumes that the effects of density-dependence are influenced by the numbers of individuals as some unspecified point in the life cycle. Even organisms that reproduce at discrete time intervals, either naturally or through artificial manipulation in the laboratory, will often have different life stages in which the effects of density are different. The impact on population dynamics of density-dependence in preadult life stages has been carefully described by Prout (1980) and Prout & McChesney (1985). In population genetics, the application of simple models for the estimation of fitness need to take account of the preadult life stages if meaningful estimates are desired (Prout 1965, 1971*a*, *b*). The lesson from these studies is that the application of simple models, even to relatively simple laboratory populations, may ignore certain important life-historical details of the organism which will compromise the ability to test the theory.

(ii) *Organism-specific models*

Recently, theoretical work in population biology has focussed on models that take into account species-specific details. Models of plant population dynamics have been developed to take into account close neighbour competition, which is a peculiar feature of sessile plants (Pacala & Silander 1985). Likewise, populations of intertidal invertebrates are often space limited, and their population dynamics require different sorts of models (Roughgarden *et al.* 1985).

I have attempted to develop models of *Drosophila* population dynamics which take into account the action of density on the various life stages of *Drosophila* (Mueller 1988*b*). This theory is outlined in figure 2. High larval density affects the amount of food available to the larval population and the amount of food consumed by each larva is assumed to have a normal

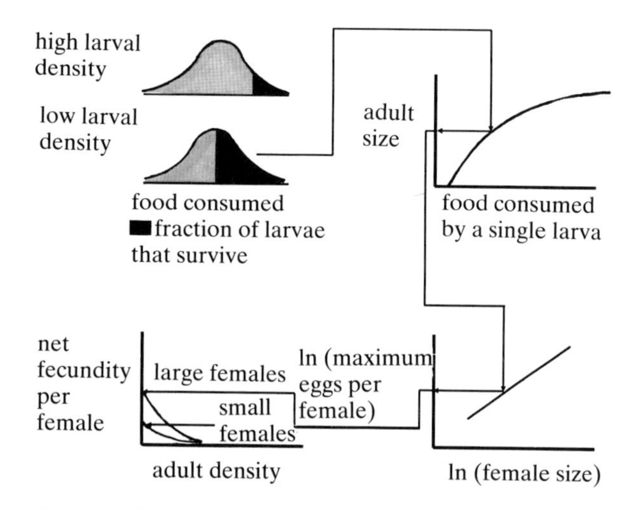

Figure 2. A graphical summary of the population dynamic model of Mueller (1988*b*). See text for details.

distribution. To survive successfully, a larva must consume a minimum amount of food. Survivors are shown as the dark portion of the normal distribution in figure 2. Adult size is determined by the amount of food consumed as a larva (figure 2). This is important for population dynamics of *Drosophila* because larger females lay more eggs than smaller females. Even very small females may lay 20–30 eggs per day (Mueller 1987) which is incompatible with stable population dynamics (Mueller 1988*b*). Net female fecundity is also modulated by adult density and may decline dramatically with increasing adult numbers (figure 2). The form of the response of female fecundity to adult density is crucial for the ultimate stabilization of population dynamics about the carrying capacity. Recently, we have seen that the shape of these curves is quite sensitive to the nutritional state of the adults and, consequently, the stability characteristics of the population may be sensitive to the levels of food available to adults (L. D. Mueller, unpublished results).

The ability to understand the evolution of life history, even in carefully controlled environments, will depend not only on carefully controlled experiments but also on the development of theory that incorporates the relevant ecological details of the experimental organism in the study environments.

I thank R. Southwood and an anonymous referee for comments on this work and for support NIH grant BRSG S07 RR07008 from the Biomedical Research Support Grant Program, Division of Research Resources.

REFERENCES

Bakker, K. 1961 An analysis of factors which determine success in competition for food among larvae in *Drosophila melanogaster*. *Arch. Neerl. Zool.* **14**, 200–281.

Bierbaum, T. J., Mueller, L. D. & Ayala, F. J. 1989 Density-dependent evolution of life-history traits in *Drosophila melanogaster*. *Evolution* **43**, 382–392.

Botella, L. M., Moya, A., González, C. & Ménsua, J. L. 1985 Larval stop, delayed development and survival in overcrowded cultures of *Drosophila melanogaster*: effect of urea and uric acid. *J. Insect Physiol.* **31**, 179–185.

Boyce, M. S. 1984 Restitution of *r*- and *K*-selection as a model of density-dependent natural selection. *A. Rev. Ecol. Syst.* **15**, 427–447.
Burnet, B., Sewell, D. & Bos, M. 1977 Genetic analysis of larval feeding behavior in *Drosophila melanogaster* II. Growth relations and competition between selected lines. *Genet. Res. Camb.* **30**, 149–161.
Charlesworth, B. 1980 *Evolution in age-structured populations.* Cambridge University Press.
Gadgil, M. & Solbrig, O. T. 1972 The concept of *r*- and *K*-selection: Evidence from wild flowers and some theoretical considerations. *Am. Nat.* **106**, 14–31.
Joshi, A. & Mueller, L. D. 1989 Evolution of higher feeding rate in *Drosophila* due to density-dependent natural selection. *Evolution* **42**, 1090–1093.
Joshi, A. & Mueller, L. D. 1991 Directional and stabilizing density-dependent natural selection for pupation height in *Drosophila melanogaster.* (Submitted.)
MacArthur, R. H. 1962 Some generalized theorems of natural selection. *Proc. natn. Acad. Sci. U.S.A.* **48**, 1893–1897.
MacArthur, R. H. & Wilson, E. O. 1967 *The theory of island biogeography.* Princeton University Press.
McNaughton, S. J. 1975 *r*- and *K*-selection in *Typha. Am. Nat.* **109**, 251–261.
Mueller, L. D. 1987 Evolution of accelerated senescence in laboratory populations of *Drosophila. Proc. natn. Acad. Sci. U.S.A.* **84**, 1974–1977.
Mueller, L. D. 1988*a* Evolution of competitive ability in *Drosophila* by density-dependent natural selection. *Proc. natn. Acad. Sci. U.S.A.* **85**, 4383–4386.
Mueller, L. D. 1988*b* Density-dependent population growth and natural selection in food-limited environments: the *Drosophila* model. *Am. Nat.* **132**, 786–809.
Mueller, L. D. 1990 Density-dependent natural selection does not increase efficiency. *Evol. Ecol.* **4**, 290–297.
Mueller, L. D. & Ayala, F. J. 1981 Trade-off between *r*-selection and *K*-selection in *Drosophila* populations. *Proc. natn. Acad. Sci. U.S.A.* **78**, 1303–1305.
Mueller, L. D. & Sweet, V. F. 1986 Density-dependent natural selection in *Drosophila*: evolution of pupation height. *Evolution* **40**, 1354–1356.
Pacala, S. W. & Silander, J. A. 1985 Neighborhood models of plant population dynamics. I. Single-species models of annuals. *Am. Nat* **125**, 385–411.
Pianka, E. 1970 On *r*- and *K*-selection. *Am. Nat.* **104**, 592–596.
Pianka, E. 1972 *r*- and *K*-selection or *b* and *d* selection? *Am. Nat.* **106**, 581–588.
Prout, T. 1965 The estimation of fitness from genotypic frequencies. *Evolution* **19**, 546–551.
Prout, T. 1971*a* The relation between fitness components and population prediction in *Drosophila*. I. The estimation of fitness components. *Genetics* **68**, 127–149.
Prout, T. 1971*b* The relation between fitness components and population prediction in *Drosophila*. II. Population prediction. *Genetics* **68**, 151–167.
Prout, T. 1980 Some relationships between density-independent selection and density-dependent population growth. *Evol. Biol.* **13**, 1–68.
Prout, T. & McChesney, F. 1985 Competition among immatures affects their adult fertility: population dynamics. *Am. Nat.* **126**, 521–558.
Robertson, F. W. 1957 Studies in quantitative inheritance. XI. Genetic and environmental correlation between body size and egg production in *Drosophila melanogaster. J. Genet.* **55**, 428–443.
Roughgarden, J. 1971 Density dependent natural selection. *Ecology* **52**, 453–468.
Roughgarden, J., Iwasa, Y. & Baxter, C. 1985 Demographic theory for an open marine population with space limited recruitment. *Ecology* **66**, 54–67.
Sewell, D., Burnet, B. & Connolly, K. 1975 Genetic analysis of larval feeding behavior in *Drosophila melanogaster. Genet. Res. Camb.* **24**, 163–173.
Slansky, F. & Feeny, P. 1977 Stabilization of the rate of nitrogen accumulation by larvae of the cabbage butterfly on wild and cultivated food plants. *Ecol. Monogr.* **47**, 209–228.

Discussion

T. R. E. Southwood (*University of Oxford, Oxford OX1 3PS, U.K.*). Professor Mueller's results on the food requirement for larvae did not show any clear picture between the different régimes. Has he corrected the amounts eaten per larva to allow for the different masses of the adults, those in the *K* régime being larger? If he did this he might find that the *K* régime individuals produced more milligrams of adult per milligram of food, that is, they would be more efficient, which intuitively one might expect.

L. D. Mueller. The question is well posed and its answer requires some clarification of terms and experiments. The definition of efficiency in my studies has been the minimum amount of food a larva must consume to successfully pupate. By studying the survival of larvae placed individually in vials with measured amounts of food it was shown (Mueller 1990) that the *r*-populations were in fact more efficient than the *K*-populations. At these very low levels of food the adults that survive are very small and there is no indication that the minimum size of adults differs between the *r*- or *K*-populations. However, as Professor Southwood has suggested, if a *K*-larva consumes slightly more than the minimum amount of food it may be more efficient at turning the food into biomass.

From the experiments I have done on groups of larvae that have been given measured amounts of food the growth curves of individual larvae can be estimated. Figure d1 illustrates these sorts of curves for the K_1 population and its control (r_1), and the K_3 population and its control (r_3). It appears that a K_1 larva can become a larger adult at most intermediate food levels. However, this apparent advantage is not seen with the K_3 and r_3 data. The curves for the K_2 and r_2 populations are, likewise, nearly identical. Thus, analysis of these data give no indication of an increase in efficiency (as this term is used by Professor Southwood) of the *K*-larvae.

It is also true that when many larvae are crowded in cultures with standard food that the viability and average

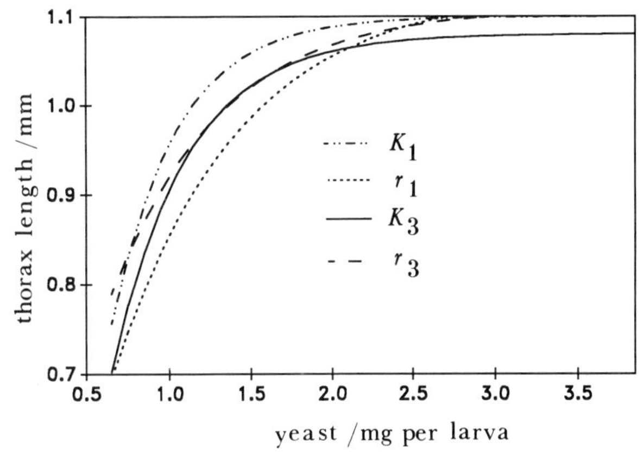

Figure d1. Growth curves of *r*-larvae and *K*-larvae given measured amounts of food. See text for details.

adult size of the surviving K-adults is greater that the r-adults (Bierbaum *et al.* 1989). However, in these crowded cultures it is difficult to say that the size differences observed show differences in efficiency. In these crowded cultures survival and growth is also affected by waste products that are consumed by the larvae (this is not a problem in the experiments that gave rise to figure 1 as the total numbers of larvae were low and constant). Some recent studies in my laboratory have shown that the K-populations tolerate high urea levels much better than do the r-populations. Consequently, it remains a possibility that the size differences observed by Bierbaum *et al.* are a result of differential tolerance to wastes and not efficiency of food conversion.

Comparing life histories using phylogenies

PAUL H. HARVEY AND ANNE E. KEYMER

Department of Zoology, University of Oxford, South Parks Road, Oxford OX1 3PS, U.K.

SUMMARY

The comparative method as recently developed can be used to identify statistically independent instances of life-history evolution. When life-history traits show evidence for correlated evolutionary change with each other or with ecological differences, it is often possible to single out the trade-offs and selective forces responsible for the evolution of life-history diversity. Suites of life-history characters often evolve in concert, and recent optimality models incorporating few variables show promise for interpreting that evolution in terms of few selective forces. Because hosts provide well-defined environments for their parasites, when host–parasite phylogenies are congruent it is possible to test ideas about the evolution of particular life-history and size-related traits.

1. INTRODUCTION

Life histories have evolved to differ among species. In this paper we ask how those differences should be examined if we are to make sense of why they evolved. The answer to our question involves two steps. First, we shall describe how comparisons across species are used in evolutionary biology. Second, we shall use the life-history theory described by previous contributors as a basis for deciding which comparisons are likely to prove the most useful for distinguishing among conflicting ideas. Throughout, we shall illustrate our points with examples based on life-history variation drawn from various animal groups.

2. THE COMPARATIVE METHOD IN EVOLUTIONARY BIOLOGY

Like astronomy or geology, evolutionary biology is largely a science about history. Evolutionary biologists must, therefore, use methods that are appropriate for retracing history. The role of experimentation is limited for several reasons, not least of which are the timescales involved. Similarly, observations of evolution in action are both scanty and likely to be of very limited relevance to major questions about the diversity of life on earth. The comparative method, however, provides a means for testing many ideas about why evolution has taken particular routes. It also provides a natural way of cataloguing diverse data sets.

The important processes that make the comparative method work are parallel and convergent evolution: when similar evolutionary responses occur to similar selective pressures, then evolutionary biologists are able to identify independently evolved correlations between characters and environments. Indeed, whole clusters of characters are likely to show correlated evolutionary change, so a successful comparative analysis can have considerable explanatory power. For example, Turner (1975) pointed out that the association of certain butterfly caterpillars with toxic food plants was, as expected, linked with an ability to sequester toxins (and consequent avoidance by predators), the evolution of aposematic or warning coloration, small home range sizes and ranges of courtship, communal roosting, delayed sexual maturity, and increased longevity. From a simple change in diet, associated responses in morphology, behaviour, and life history seem to follow in this one taxonomic group.

As mentioned above, the logic of the comparative method rests on the processes of parallel and convergent evolution. It is, therefore, important to recognize that similarity among extant species may be inherited from common ancestors, rather than having evolved by parallel or convergent evolution. (There are many reasons why closely related species are likely to remain similar: speciation into similar niches and phylogenetic time lags are two. These and others are discussed by Harvey & Pagel (1991).) Accordingly, species values do not provide independent data for comparative analyses, and the best comparative tests must utilize available material on phylogenetic relatedness among the species in a sample. Take the phylogenetic tree in figure 1 in which the higher nodes denote generic status, the next lower nodes family status, and the bottom node shows that all species belong to the same order. Species within genera are very similar to each other on characters X and Y. However, species within the same family but belonging to different genera are more similar to each other than species from different families. This means that neither family nor generic averages provide suitable data points for comparative analyses. How, then, should such analyses be performed?

Two types of procedure are available, which we call

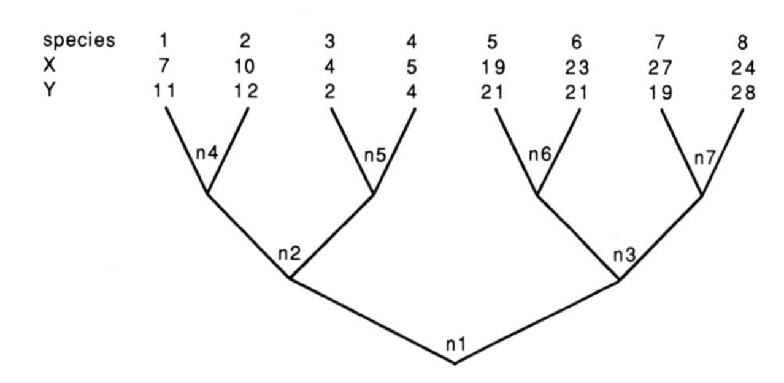

Figure 1. A hypothetical phylogenetic tree showing the relation between eight extant species. Pairs of species emanating from each of the four higher nodes in the figure (n4, n5, n6, n7) are from the same genus, and pairs of genera arising from the intermediate level nodes (n2, n3) are from the same family. Characters X and Y differ among the species, but species with more recent common ancestors have more similar values for both X and Y.

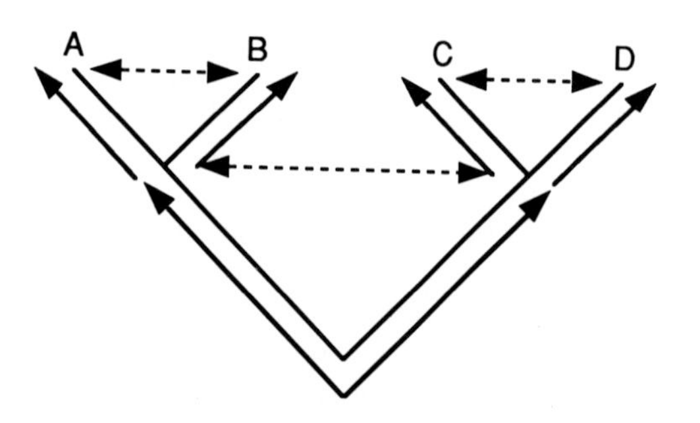

Figure 2. A hypothetical phylogenetic tree showing the relation between four species. The six full-arrowed lines denote directional comparisons that are possible between ancestors and descendants. The three dotted-arrowed lines denote non-directional comparisons. If the phylogenetic tree is known but character states are available for only extant species, there would be only three degrees of freedom in the data (the number of species minus one) and non-directional analyses are appropriate. If ancestral character states for the nodal species are also known there will be six degrees of freedom in the data (the number of species plus the number of nodes minus one), thus allowing directional analyses to be used.

'directional' and 'non-directional' analyses. For the phylogenetic tree shown in figure 2, the full arrows denote directional analyses which involve comparisons between seven different ancestral and descendant taxa. If ancestral phenotypes are known, such directional analyses are satisfactory – the number of nodes plus extant species is seven, allowing six comparisons to account for all the degrees of freedom in the data. Unfortunately, ancestral phenotypes are usually not known and with data for only four extant species being available, three independent non-directional comparisons are possible. Felsenstein (1985) points out that, under a Brownian motion model of evolutionary change, the three non-directional comparisons labelled by dotted lines are statistically independent. The dotted line between the two nodes is a comparison between generic values, calculated as the averages of constituent species values. The non-directional comparisons have a ready evolutionary interpretation: they denote differences that have evolved between daughter taxa emanating from a common ancestor.

Grafen (1989) and Pagel & Harvey (1989) have described methods for applying Felsenstein's non-directional comparative technique to real data. Their methods are designed to overcome a number of problems, including how to deal with multiple nodes (many genera consist of three or more species which presumably mask two or more dichotomies) and how to determine branch lengths from real data. Here we sweep aside such details and summarize the procedures for a primitive non-directional comparative test in figure 3. As is evident from the figure, ultimately a series of contrasts for one variable, perhaps a life-history variable, is compared with a series of contrasts for another variable, perhaps a measure of lifestyle or ecology. Usually, larger contrasts will be associated with lower nodes – differences between families in an order are likely to be larger than those between species within a genus. Evidence for correlated evolution is provided by contrasts for one character not changing independently of those for another character. It is always important to examine scatterplots of contrasts to check for instances of nonlinear correlated evolution (see Harvey & Pagel 1991).

(*a*) *An example of a non-directional comparative analysis*

We present here a summary of a typical non-directional comparative test of a theory about life-history differences. It has often been suggested that species with high metabolic rates for their body sizes live faster and shorter lives: that is, they have higher fecundity (measured in terms of number of young produced per unit time) and shorter lifespans (see, for example, Calder & King 1974; Fenchel 1974; Millar 1977; Blueweiss *et al.* 1978; Sacher 1978; McNab 1980, 1986; Western & Ssemakula 1982; Swihart 1984). Avian phylogenetic relations are now reasonably well known, owing in large part to the work of Sibley and his colleagues using DNA–DNA hybridization methods (see Sibley *et al.* 1988). Furthermore, avian basal metabolic rates and life histories are also widely described in the literature. Now that the data are available, it is possible to test whether the metabolic rate theory holds.

Trevelyan *et al.* (1990) used a non-directional independent comparisons method to determine whether there was evidence for correlated evolution between metabolic rate and life histories across a sample of 325 bird species when body size was held constant. Eleven life-history variables were used in the analysis: hatchling mass, egg mass, clutch size, incubation time, age at fledging, age at independence, age at first breeding, lifespan, number of broods per year, interval between eggs, and clutch mass. Three sets of contrasts were calculated: one for the life-history variable of interest, one for body mass, and one for metabolic rate. There was clear evidence for correlated

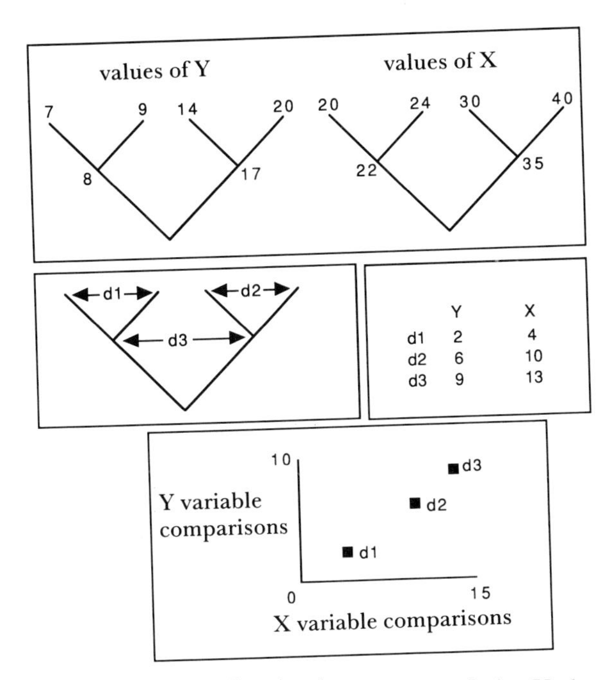

Figure 3. A non-directional contrast analysis. Under a Brownian motion model of evolution, d1, d2 and d3 provide independent comparisons (Felsenstein 1985). In the example, positive correlated evolutionary change between the X and Y variables is evident: large positive change in X is associated with large positive change in Y, while a small positive change in X is associated with a small positive change in Y. Each of the three comparisons in X is conditioned to be positive.

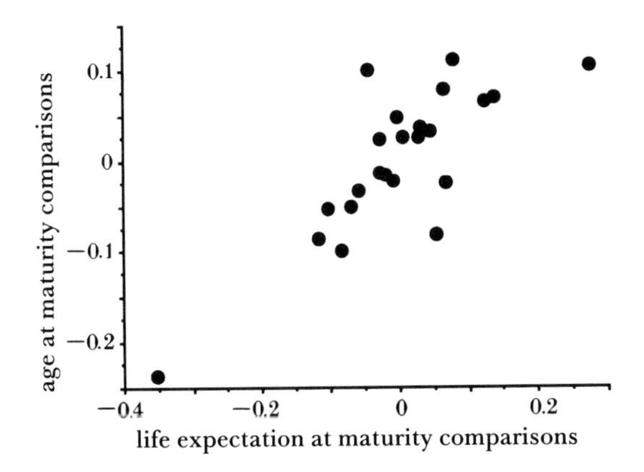

Figure 4. Independent comparisons calculated from Millar & Zammuto's (1983) data. Age at maturity and life expectation comparisons are both corrected for body mass in the figure, being given as the residuals from the regressions of each set of comparisons on body mass comparisons. The correlation is highly significant ($r = 0.80$, $n = 23$, $p < 0.001$). The bottom left point is a comparison between *Ochotona princeps* and *Sylvilagus floridanus*, and when this outlier is removed the correlation remains highly significant ($r = 0.66$, $n = 22$, $p < 0.001$). After Harvey & Pagel (1991).

evolution between all three variables, but partial correlation analysis on the contrasts showed that life-history variation was not usually significantly correlated with variation in metabolic rate when body mass was held constant: of eleven life-history variables tested, only one partial correlation was significant at the 5% level. However, life-history variation was significantly correlated with variation in body mass when variation in metabolic rate was held constant: nine of the eleven partial correlations were significant at the 5% and three at the 1% level. It seems, then, that metabolic rate differences might not play such a central role in avian life-history variation as had been thought. Similar analyses on mammal life histories, metabolic rates and body masses lead to similar conclusions (Harvey *et al.* 1991).

3. ASKING THE RIGHT QUESTIONS

(*a*) *Identifying trade-offs*

The key to understanding life-history variation is provided by identifying correctly the trade-offs that prevent organisms from being what Law (1979) has called 'Darwinian Demons'. If trade-offs were not involved, each species would start producing copious offspring at birth and continue doing that through to eternity. In practice, there are trade-offs between components of fecundity, between components of mortality, and between fecundity and mortality. Different environments select for different optimal life-histories, and Southwood (1977, 1988) has prescribed an agenda that seeks natural ways of classifying habitats according to certain templets that are the key determinants influencing changes in optimal life-histories.

If they are to start with basic questions, comparative studies might be used to identify those trade-offs that are important in life-history evolution. It has often been pointed out (see Partridge & Harvey 1985, 1988) that comparative data can fail to identify trade-offs that exist, or even suggest positive associations between the component variables of a real trade-off. For example, imagine that there is a trade-off between clutch size and survival in a temperate passerine bird species such that if an individual female lays a larger clutch she is less likely to survive to breed again. Even when such trade-offs exist within species, comparative tests can show positive correlations: the fittest birds in a population may lay large clutches and have a higher chance of overwinter survival. However, this does not mean that cross-taxonomic comparative evidence for trade-offs is not with pursuing.

Recently, Blackburn (1991) has reported the first comparative study to use a non-directional comparative analysis for identifying a trade-off in life-history evolution. Blackburn's analysis is of particular interest because it tests for one of the classic trade-offs in life-history studies, that between egg size and clutch size in birds. Lack (1967) suggested that individual female birds could allocate their finite resources to producing few larger eggs, or more smaller eggs. Given that many large eggs will produce a greater number of more viable offspring, Lack was hypothesizing a trade-off between egg size and clutch size. Lack's idea was likely to be particularly appropriate to precocial taxa in which the mother feeds the offspring little if at all, so nutrient limitation during egg production would limit fecundity. Accordingly, Lack performed a comparative test of his idea using the waterfowl (Anatidae), and found the predicted negative relation between clutch size and egg size.

Rohwer (1988) questioned Lack's conclusions, partly on methodological grounds but also in the light of more reliable data that had recently become available. Larger females are likely to have more resources to invest in their offspring, and so Rohwer controlled statistically for body size in his analysis. He found that the variance in clutch size accounted for only 13% of the variance in egg mass when female body size was controlled for.

Blackburn pointed out that while Rohwer's data were better than Lack's, his method of analysis remained primitive and may have led to the wrong conclusions. Residuals from a major axis line were used by Rohwer to control for the effects of body mass, which is an incorrect procedure because those deviations (unlike deviations from a Model I regression line) were correlated with body mass. Furthermore, taxonomic association was not properly controlled for in Rohwer's study, which meant that some speciose taxa (e.g. 42 of the 151 species belong to the genus *Anas*) could influence the results. Blackburn calculated contrasts using Rohwer's data in a non-directional phylogenetic comparative analysis of the sort described above. When partial correlation analysis was done on the contrasts, clutch size accounted for 29% of the variance in egg mass after controlling for female body size. Read & Harvey (1989) found similar evidence for a trade-off between litter size and neonatal mass in mammals, independent of maternal body mass. Partridge & Sibly (this symposium) mention the importance of controlling for differences in other life-history characters and for differences in habitat when seeking to identify trade-offs. When Blackburn (1990) performed such tests, the negative correlation between contrasts for egg mass and clutch size was strengthened. However, even though such comparative studies leave little room for doubt about the presence of particular trade-offs, we agree with Partridge & Sibly that they do not allow the construction of accurate trade-off curves.

The central role of trade-offs in life-history theory means that comparative patterns must often be interpreted in terms of appropriate trade-offs, and the interpretations subject to test. An example is provided by work on age at maturity among female mammals. It has often been argued that age at maturity is allometrically related to body mass, that is it relates to body size according to a power function, so that a double-logarithmic plot reveals a linear relation between age at maturity and adult body mass. Such allometric relations require explanation. For example, larger animals take longer to grow to adult size and so will mature at a later age. However, for their body size, some mammals mature at a later age than others. Why should this be? Harvey & Zammuto (1985) pointed out that mammals with low mortality rates for their body mass were the ones with late ages at maturity. Harvey & Zammuto's analysis, and a subsequent modification by Sutherland *et al.* (1986) which plotted age at maturity against life expectation at maturity, were based on a cross genus comparison. Such analyses, as we pointed out in the discussion above relating to figure 1, are not fully appropriate. However, when the analysis is repeated using a non-directional contrasts analysis, the pattern is still evident (see figure 4).

It is a matter of demographic necessity that birth rates must ultimately match death rates if populations are to persist at demographic equilibrium, with the consequence that some components of fecundity will match some components of mortality (Sutherland *et al.* 1986). However, this will not explain why some species apparently delay reproduction beyond an age when they are fully grown. In particular, these ages at maturity, while varying a little with state of nutrition, do not change sufficiently in well fed zoo populations to account for the relation in figure 4. Instead, it has been argued, we might invoke a trade-off to explain such results (Ashmole 1963; Charlesworth 1980): if there is a cost to reproduction but the efficiency of reproduction increases with age, then species with low rates of mortality will have been selected to delay the onset of reproduction (Harvey *et al.* 1989*b*). For example, consider a species of mammal that will die after producing young (such semelparous species are known). If they can produce two offspring as two-year olds or ten offspring as three-year olds, they will delay reproducing until they are three so long as adult mortality is sufficiently low. If the chance of surviving from age two to three is just 1%, the mammal will have been selected to breed as a two-year old. Here the trade-off is between delaying reproduction to a time when reproductive efficiency is high and dying while waiting.

The explanation we have just given for the relation in figure 4 involved an increase in reproductive efficiency with age: producing offspring should cost older (but not senescent) animals less, or older mothers should produce more offspring for the same cost. Is there evidence for this? Indeed, there is circumstantial evidence – in various species, older mothers produce larger offspring, larger litters, young with higher rates of survival, and have shorter interbirth intervals (Promislow & Harvey 1990). However, some or all of these patterns may result from experience of older parents, and experiments will be necessary to distinguish age from experience. It is possible that such experiments have been reported in the agricultural literature.

(*b*) *Comprehensive models*

Simple relations, such as that between age at maturity and rate of mortality, may be explained in terms of simple trade-offs, but the comparative method in life history is providing evidence for suites of correlated life-history characters. For example, Read & Harvey (1989) found that several measures of increased fecundity covaried across orders of mammals: short gestation lengths, early ages at weaning, and short periods as independent juveniles before maturation were highly correlated with each other independent of adult body mass. Similar patterns have been found in birds (Bennett & Harvey 1988; Saether 1988). It is possible to produce *ad hoc* but testable theories to explain many or all of these different correlations (for examples, see Harvey *et al.* 1989*b*). Perhaps more exciting, however, is the prospect of producing integrated optimality models of life-history evolution that incorporate known functional relations between growth rates, body mass, mortality rates, and various life-history parameters. Such models inevitably incorporate both constraints and trade-offs. Charnov (1991) has recently made a start in this direction in a model of mammalian life-history evolution which predicts many of the interspecific relations that we actually find in nature. We consider Charnov's example in a little detail, attempting a purely verbal exposition because we are less concerned with the precise numerical values and functional relations involved than we are with his overall approach.

Charnov takes as his starting point the assumption that mammals grow according to a fixed growth law, which dictates that the amount of weight put on per unit time increases with time and, therefore, with weight. In other words, bigger mammals grow faster. Charnov assumes that mammals cease growing when they mature, and the energy that could have been channelled into growth is diverted into offspring production. This means that larger mammals have more energy to put into offspring production per unit time. However, larger-bodied mammals produce larger offspring, with the net result that their fecundity is lower. Now, mammals are selected to maximize their lifetime reproductive success, which depends on fecundity, adult mortality, and juvenile mortality. Charnov views adult mortality as extrinsic to the system, with natural selection acting on the age at maturity to maximize lifetime reproductive success. Increasing the age of maturity increases size and, thereby, reduces fecundity. It also prolongs the period of immaturity and decreases the probability of a juvenile surviving to reproduce. Charnov considers that mortality rates plateau well before maturity so that adult mortality rates are not influenced by changes in the age at maturity. In contrast, altering adult mortality rate changes the optimal age at maturity (set as the age that maximizes lifetime reproductive success), thereby dictating adult body mass. For a population in demographic equilibrium, of course, the average number of daughters produced during a female's lifetime is one. Charnov considers that average juvenile mortality adjusts so that this equality holds.

Some interesting factors emerge from Charnov's formulation. First, for animals with similar growth curves and for which adult body mass determines body mass at weaning, fecundity (number of offspring produced by a female per unit time) is considered to vary simply with the efficiency with which females can transfer units of potential growth into units of offspring. Second, adult mortality rate is extrinsic to the system and is not influenced by adult body mass. In fact the reverse is true, with adult mortality rate influencing adult body mass through its effect on the optimal age at maturity. Third, as we can describe age at maturity in terms of adult body mass (because of the universal growth law), any variable that can be defined in terms of adult body mass can equally well be defined in terms of age at maturity. For example, the growth relation can be used to eliminate adult body size from equations describing fecundity, adult mortality rate, and average immature mortality producing dimensionless numbers (the topic of Charnov's paper in this symposium). Fourth, since age at maturity, adult mortality rate, fecundity and average immature mortality can each be expressed in terms of adult body mass, we can examine the relation between those variables when adult mass is held constant.

The relations between the four variables (age at maturity, pre-adult mortality, adult mortality, fecundity) are defined precisely so long as the various assumptions of Charnov's model hold. Let us review some of those assumptions: there is a fixed growth law, juvenile mass is defined precisely by adult mass, and the efficiency of transfer from units of parental growth into units of young is fixed. Each of these assumptions gives rise to a constant in Charnov's equations. If those constants are changed in value, so are age at maturity, adult body mass, fecundity, and average immature mortality. What is important for the purposes of this article is that the assumptions will not quite hold, so that the constants differ a little among species. If they differ among species in an uncorrelated fashion (e.g. changes in the growth law are not accompanied by predictable changes in the efficiency of transfer of energy from adult growth to production of young) then Charnov predicts the various size-independent correlations between life-history variables and between mortality and fecundity found by Harvey & Zammuto (1985), Sutherland *et al.* (1986), Read & Harvey (1989), and Promislow & Harvey (1990).

Let us return to our example of the positive relation between early age at maturity and high adult mortality. Can we see why in Charnov's model this relation should hold independently of differences in body mass? The way that we have described the model, fecundity is higher for smaller mammals because the extra energy for reproduction that is available to larger mammals is more than offset by the increased size of the weaned offspring that they produce. However, when Charnov applies optimality theory to find the age at maturity, he assumes that mass at weaning is fixed. Within that portion of his model, therefore, larger animals can produce more offspring per unit time if they mature later (i.e. at a greater mass). In short, reproductive efficiency increases with age at maturity in Charnov's model. The trade-off, then, is between increasing reproductive output by delaying reproduction, and dying while waiting. A useful component of Charnov's formulation, therefore, is that by specifying relations precisely it allows us to identify relevant trade-offs. Interestingly, the trade-off implicit in Charnov's model is not the same as that suggested earlier in this paper. The previous model assumed increased reproductive efficiency with age per se, and a cost to reproduction in terms of continued adult survival. Charnov makes neither assumption, but he does assume increased efficiency of reproduction with an increased age at maturity and, for his

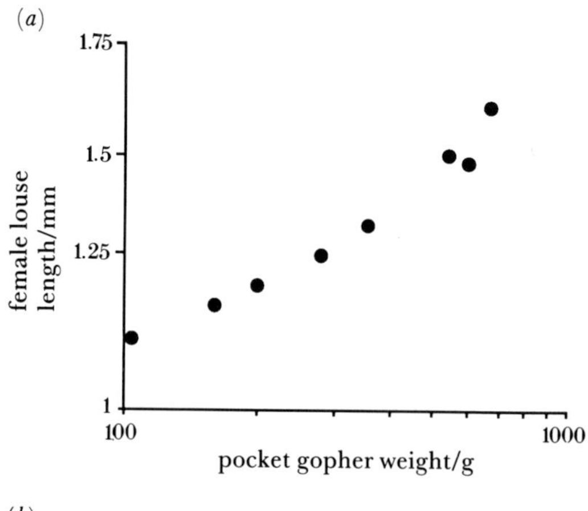

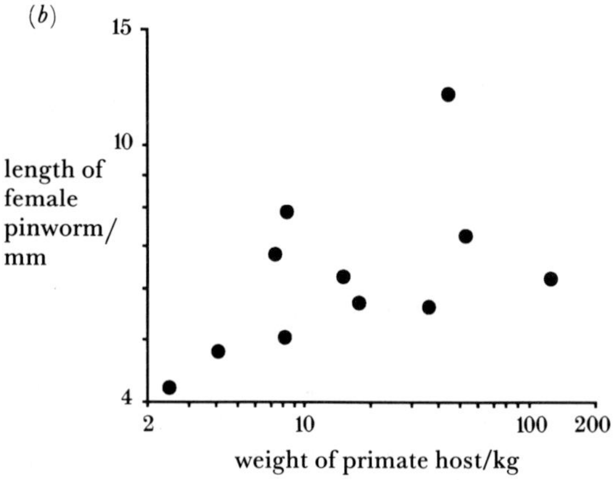

Figure 5. Size relations across species between (*a*) pocket gophers and their associated chewing lice, and (*b*) between primates and their pinworms.

optimization, that all phenotypes wean offspring of the same mass.

(*c*) *From correlated evolution to coevolution*

So far in this article we have focussed on the use of comparative methods to show correlated evolution between life-history traits. As we have mentioned, and as Southwood and others have emphasized, different habitats probably favour the evolution of different lifestyles. We saw one example with Turner's butterflies where association with toxic plants favours the evolution of a long lifespan and delayed maturity. One case where convergent evolution of life-history traits has been identified is in primates. Ross (1987) found that primate species living in tropical rain forests lived much slower lives for their body sizes than those in secondary forests, riverine forests or savannah. Harvey *et al.* (1989*a*) pointed out that the same trend was found in five of six different families in the sample and, therefore, had evolved on separate occasions. Alternative explanations invoking r/K selection (Ross 1987) and differential mortality (Harvey *et al.* 1989*a*) have been suggested for the pattern. Such examples of the correlated evolution of life-history traits with differences in environments are, however, not common as has been discussed elsewhere (see, for example, Partridge & Harvey 1988; Harvey *et al.* 1989*b*).

In addition to studying correlated evolution, the new comparative techniques discussed in this paper open up the unexplored possibility of studying coevolution between life-history traits, where 'the study of coevolution is the analysis of reciprocal genetic changes that might be expected to occur in two or more ecologically interacting species and the analysis of whether the expected changes are actually realized' (Futuyma & Slatkin 1983, pp. 2–3). Our aim is to understand the reasons for the evolution of life-history differences. That means defining the selective forces responsible for those differences. It is usually impractical to define fine-grained components of environmental diversity that might be responsible for the evolution of life-history differences among closely related species. Some host–parasite relations might provide an exception. Closely related hosts can have markedly different life histories, which should impose life-history differences on their parasites. The host constitutes a predictable and homeostatically controlled environment for a parasite, and those aspects of the host environment most likely to affect parasite life history (such as resource availability and temporal stability) are well characterized by the life history of the host. In turn, parasites invariably influence their host's fitness. Accordingly, when parasite phylogeny is congruent with host phylogeny it should be possible to use the methods of independent comparisons to examine coevolution, although cause and effect can only by hypothesized in the absence of information about ancestral character states.

To provide simple illustrations of the potential use of such methods, we sought examples of host-specific parasites that had speciated with their hosts, so that we had congruent host–parasite evolutionary trees. Of course, no taxa fit these criteria completely: most parasites have more than one host species, and among the specialists there is usually clear evidence for switching among host lineages. Even for the most appropriate taxa, the data on parasite life histories are extremely limited. We choose two examples: chewing lice and roundworms. Specifically, we concentrate on lice of the mallophagan family Trichodectidae which infect pocket gophers in the rodent family Geomyidae (Hafner & Nadler 1988), and pinworms of the genus *Enterobius* found in primates (Brooks & Glen 1982). Life-history data for the mammals were extracted from sources given by Read & Harvey (1989) and supplemented by information on pocket gopher life-histories supplied to us by Dr Mark Hafner. Life history data for the parasites were taken from a variety of published sources (Cameron 1929; Sandosham 1950; Skrjabin *et al.* 1960; Yen 1973; Quentin *et al.* 1979; Price & Hellenthal 1980, 1981; Timm & Price 1980; Hugot & Tourte-Schaefer 1985; Hugot 1987; Hellenthal & Price 1989*a*, *b*).

Simple cross-species analyses reveal clear relations (figure 5): large parasites are associated with large hosts in both parasite–host groups. But how do we control for possible phylogenetic effects in both parasite *and* host? In these examples, the phylogenies are more-or-less congruent (figure 6: the examples are for illustrative purposes only – we have ignored some evidence for host-switching in the lice, while for

(*a*)

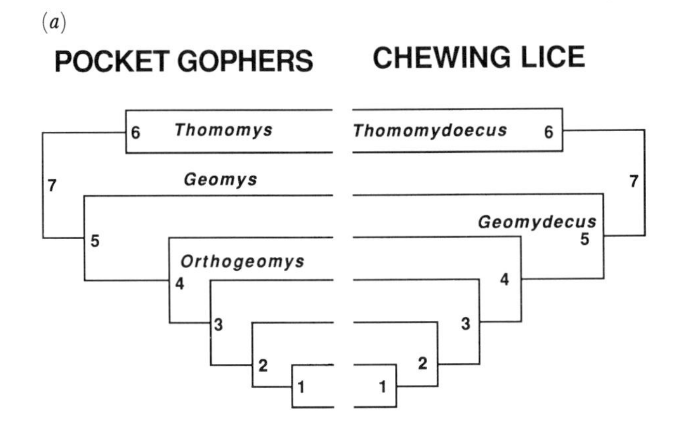

(*b*)

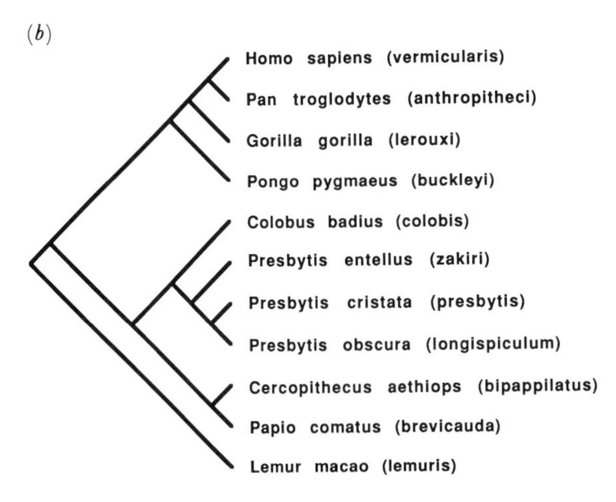

Figure 6. Congruent phylogenetic trees of (*a*) pocket gophers and their chewing lice, and of (*b*) primates and their *Enterobius* pinworms. Uncertainties of tree structure have, for the purposes of illustration, been decided in favour of congruence in each case. References for tree reconstruction are given in the text.

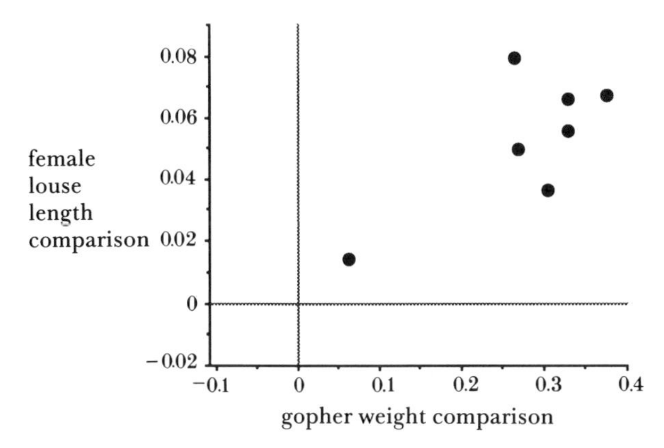

Figure 7. Independent evolutionary non-directional comparisons of chewing louse length on gopher host adult body weight. In each case, increased host weight is associated with increased louse length (all points are in the upper right-hand quadrant).

Enterobius our parasite phylogeny is derived in part from the host one), so independent comparisons were made in exactly the manner already described (see figure 3). Life-history data were not readily available

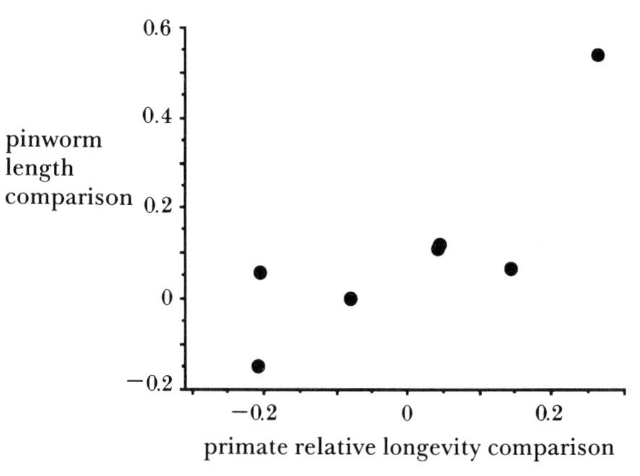

Figure 8. Independent evolutionary comparisons between pinworm length and primate longevity corrected for body size ($r = 0.81$, $n = 7$, $p < 0.05$).

for pocket gophers or their associated lice but, in an independent comparisons test, we can see that the evolution of increased body size in the host is invariably associated with increased size of parasites (figure 7). This is, perhaps, a surprising finding because we should not have expected host size per se to set a constraint on the size of its lice. However, it is generally the case that larger female invertebrates are more fecund (see, for example, Sibly & Calow 1986). Perhaps the large size of the lice is really a consequence of the longer lives of their hosts – lice living on larger hosts have longer to grow before the host perishes. The trade-off, then, would be between increased fecundity with larger size and increased mortality while waiting to attain that size. The primate–pinworm association allows us one test of that idea.

Larger primate species act as hosts to larger pinworms, but primate lifespan is also correlated with pinworm size. We used an independent comparisons test to ask if the evolution of long primate lifespans when controlled for increases in primate body size was associated with increased pinworm size. Indeed it was (figure 8). The procedure for the test was to calculate three sets of non-directional comparisons: one for primate mass, one for primate lifespan, and one for pinworm size. We then regressed the comparisons for primate lifespan on those for primate mass, and plotted the residuals from that regression against the appropriate comparisons for pinworm size. Our results, therefore, support the idea that pinworm sizes evolve in response to the longevity of their hosts.

In our examples, we have assumed the simplest scenario; congruence between parasite and host phylogenies, with co-speciation and no host switching. Future analyses along these lines will probably deal with cases where parasites use more than a single host, and where parasite switching between host lineages occurs. We foresee no great technical problems in developing the appropriate statistical models, whereas the analyses could be profitable indeed because they can be used to tackle questions about coevolution between parasites and their hosts (we assumed here that host life-history determined parasite life-history), about host-specificity, and about speciation itself.

We are grateful to Professor M. Anwar, Ms A. Collie, Dr C. W. Gordon, Dr M. S. Hafner, Dr C. Lyal, Dr S. Nee, Dr L. Partridge and Mr A. Purvis for their help in the preparation of this paper. We also thank the Commission for European Communities for supporting the work.

REFERENCES

Ashmole, N. P. 1963 The regulation of numbers of oceanic birds. *Ibis* **103**b, 458–473.

Bennett, P. M. & Harvey, P. H. 1988 How fecundity balances mortality in birds. *Nature, Lond.* **333**, 216.

Blackburn, T. M. 1991 The interspecific relationship between egg size and clutch size in waterfowl – a reply to Rohwer. *Auk* (In the press.)

Blueweiss, L., Fox, H., Kudzma, V., Nakashima, D., Peters, R. H. & Sams, S. 1978 Relationships between body size and some life history parameters. *Oecologia* **37**, 257–272.

Brooks, D. R. & Glen, D. R. 1982 Pinworms and primates: a case study in coevolution. *Proc. Helminth Soc. Wash* **49**, 76–85.

Calder, W. A. & King, J. R. 1974 Thermal and caloric relations in birds. In *Avian Biology*. vol. 4 (ed. D. S. Farner & J. R. King), pp. 259–413. New York: Academic Press.

Cameron, T. W. M. 1929 The species of *Enterobius* in primates. *J. Helminth.* **24**, 171–204.

Charlesworth, B. 1980 *Evolution in age-structured populations*. Cambridge University Press.

Charnov, E. L. R. 1991 Evolution of life history variation among female mammals. *Proc. natn. Acad. Sci. U.S.A.* **88**, 1134–1137.

Felsenstein, J. 1985 Phylogenies and the comparative method. *Am. Nat.* **125**, 1–15.

Fenchel, T. 1974 Intrinsic rate of natural increase: the relationship with body size. *Oecologia* **14**, 317–326.

Futuyma, D. J. & Slatkin, M. 1983 *Coevolution*. Sunderland, Massachusetts: Sinauer Associates.

Grafen, A. 1989 The phylogenetic regression. *Phil. Trans. R. Soc. Lond.* B **326**, 119–157.

Hafner, M. S. & Nadler, S. A. 1988 Phylogenetic trees support the coevolution of parasites and their hosts. *Nature, Lond.* **332**, 258–259.

Harvey, P. H. & Pagel, M. D. 1991 *The comparative method in evolutionary biology*. Oxford University Press.

Harvey, P. H., Pagel, M. D. & Rees, J. A. 1991 Mammalian metabolism and life histories. *Am. Nat.* (In the press.)

Harvey, P. H., Promislow, D. E. L. & Read, A. F. 1989*a* Causes and correlates of life history differences among mammals. In *Comparative socioecology* (ed. R. Foley & V. Standen), pp. 305–318. Oxford: Blackwell Scientific Publications.

Harvey, P. H., Read, A. F. & Promislow, D. E. L. 1989*b* Life history variation in placental mammals: unifying the data with the theory. *Oxf. Surv. evol. Biol.* **6**, 13–31.

Harvey, P. H. & Zammuto, R. M. 1985 Patterns of mortality and age at first reproduction in natural populations of mammals. *Nature, Lond.* **315**, 319–320.

Hellenthal, R. A. & Price, R. D. 1989*a* The *Thomomydoecus wardi* complex (Mallophaga: Trichodectidae) of the pocket gopher. *J. Kansas Ent. Soc.* **62**, 245–253.

Hellenthal, R. A. & Price, R. D. 1989*b* The *Geomydoecus thomomys* complex (Mallophaga: Trichodectidae) from pocket gophers of the *Thomomys talpoides* complex (Rodentia: Geomyidae) of the United States and Canada. *Ann. ent. Soc. Am.* **82**, 286–297.

Hugot, J. P. 1987 Sur le genre *Enterobius* (Oxyuridae, Nematoda): s.g. *Colobenterobius*. I. Parasites de primates Colobinae en region ethiopienne. *Bull. Mus. Nat. d'Hist. Natur.* **A 9**, 341–352.

Hugot, J. P. & Tourte-Schaefer, C. 1985 Morphological study of two pinworms parasitic in man: *Enterobius vermicularis* and *E. gregorii*. *Ann Parasit. Hum. Comp.* **60**, 57–64.

Lack, D. 1967 The significance of clutch size in waterfowl. *Wildfowl* **19**, 67–69.

Law, R. 1979 Ecological determinants in the evolution of life histories. In *Population dynamics* (ed. R. M. Anderson, B. D. Turner & L. R. Taylor), pp. 81–103. Oxford: Blackwell Scientific Publications.

McNab, B. K. 1980 Food habits, energetics and population biology of mammals. *Am. Nat.* **116**, 106–124.

McNab, B. K. 1986 The influence of food habits on the energetics of eutherian mammals. *Ecol. Monogr.* **56**, 1–19.

Millar, J. S. 1977 Adaptive features of mammalian reproduction. *Evolution* **31**, 370–386.

Millar, J. S. & Zammuto, R. M. 1983 Life histories of mammals: an analysis of life tables. *Ecology* **64**, 631–635.

Pagel, M. D. & Harvey, P. H. 1989 Comparative methods for examining adaptation depend on evolutionary models. *Folia primat.* **53**, 203–220.

Partridge, L. & Harvey, P. H. 1985 The costs of reproduction. *Nature, Lond.* **316**, 20.

Partridge, L. & Harvey, P. H. 1988 The ecological context of life history evolution. *Science, Wash.* **241**, 1449–1455.

Price, R. D. & Hellenthal, R. A. 1980 A review of the *Geomydoecus minor* complex (Mallophaga: Trichodectidae) for *Thomomys* (Rodentia: Geomyidae). *J. med. Entomol.* **17**, 298–313.

Price, R. D. & Hellenthal, R. A. 1981 A review of the *Geomydoecus californicus* complex (Mallophaga: Trichodectidae) from *Thomomys* (Rodentia: Geomyidae). *J. med. Entomol.* **18**, 1–23.

Promislow, D. E. L. & Harvey, P. H. 1990 Living fast and dying young: a comparative analysis of life history variation among mammals. *J. Zool.* **220**, 417–437.

Quentin, J. C., Betterton, C. & Krishnasamy, M. 1979 Oxyures nouveaux ou peu connus, parasites de primates, de rongeurs et de dermopteres en Malaisie. Creation de sous-genre *Colobenerobius* n. subg. *Bull. Mus. Nat. d'Hist. Natur., Paris* **4**, 1031–1050.

Read, A. F. & Harvey, P. H. 1989 Life history differences among the eutherian radiations. *J. Zool.* **219**, 329–353.

Rohwer, F. C. 1988 Inter- and intraspecific relationships between egg size and clutch size in waterfowl. *Auk* **105**, 161–176.

Ross, C. R. 1987 The intrinsic rate of natural increase and reproductive effort in primates. *J. Zool.* **217**, 199–220.

Sacher, G. A. 1978 Longevity and ageing in vertebrate evolution. *Bioscience* **28**, 297–301.

Sæther, B. E. 1988 Evolutionary adjustment of reproductive traits to survival rates in European birds. *Nature, Lond.* **331**, 616–617.

Sandosham, A. A. 1950 On *Enterobius vermicularis* and related species. *J. Helminth.* **24**, 171–204.

Sibley, C. G., Ahlquist, J. E. & Monroe, B. L. 1988 A classification of the living birds of the world based on DNA–DNA hybridization studies. *Auk* **105**, 409–423.

Sibly, R. M. & Calow, P. 1986 *Physiological ecology of animals*. Oxford: Blackwell Scientific Publications.

Skrjabin, K. I., Shikhobalora, N. B. & Lagoovsaya, E. A. 1960 *Oxyurata of animals and man*. Moscow: Izdatelstvo Akademii Nank SSR.

Southwood, T. R. E. 1977 Habitat, the templet for ecological strategies? *J. Anim. Ecol.* **46**, 337–365.

Southwood, T. R. E. 1988 Tactics, strategies and templets. *Oikos* **52**, 3–18.

Sutherland, W. J., Grafen, A. & Harvey, P. H. 1986 Life history correlations and demography. *Nature, Lond.* **320**, 88.

Swihart, R. K. 1984 Body size, breeding season length and life history tactics of lagomorphs. *Oikos* **43**, 282–290.

Timm, R. M. & Price, R. D. 1980 The taxonomy of *Geomydoecus* (Mallophaga: Trichodectidae) from the *Geomys bursarius* complex (Rodentia: Geomyidae). *J. med. Entomol.* **17**, 126–145.

Trevelyan, R., Harvey, P. H. & Pagel, M. D. 1990 Metabolic rates and life histories in birds. *Funct. Ecol.* **4**, 135–141.

Turner, J. R. G. 1975 A tale of two butterflies. *Nat. Hist.* **84**, 28–37.

Western, D. & Ssemakula, J. 1982 Life history patterns in birds and mammals and their evolutionary interpretation. *Oecologia* **54**, 281–290.

Yen, W. C. 1973 Helminthes of birds and wild animals from Lin-Tsan Prefecture, Yunnan Province, China. II Parasitic nematodes of mammals. *Acta zool., Sinica* **19**, 354–364.

Dimensionless numbers and the assembly rules for life histories

ERIC L. CHARNOV AND DAVID BERRIGAN

Department of Biology, University of Utah, Salt Lake City, Utah 84112, U.S.A.

SUMMARY

This paper reviews recent efforts to use certain dimensionless numbers (DLNs) to classify life histories in plants and animals. These DLNs summarize the relation between growth, mortality and maturation, and several groups of animals show interesting patterns with respect to their numeric values. Finally we focus on one DLN, the product of the age of maturity and the adult instantaneous mortality, to show how evolutionary life history theory may be used to predict the value of the DLN, which differs greatly between major groups of animals.

1. INTRODUCTION

This paper is about the use of dimensionless numbers (DLNs) to characterize and classify life histories in animals and plants. DLNs are widely used in dynamical and mechanical problems where the behaviour of the system of interest often depends on the ratio or product of parameters and not the values of each alone. A simple example from theoretical biology is the equation for change in gene frequencies for a single diallelic locus under natural selection. Under the usual assumptions, the dynamics are entirely given by the relative fitnesses of the three genotypes. While actual Darwinian fitness is the product of survival and fertility (and has units of 'numbers'), relative fitness is constructed by dividing the three genotype fitnesses by the fitness of the heterozygote (and is dimensionless); thus, we get, for the genotypes AA:Aa:aa, the relative fitnesses $1\text{-}s:1:1\text{-}t$.

Mathematical functions between dimensional variables can always be rewritten in dimensionless forms by using formal techniques from so-called 'dimensional analysis' (see, for example, Giordano *et al.* (1987); Stephens (1991), for applications to behavioural ecology; or Stahl (1962) and Calder (1984) for body size and physiology). Dimensionless variables have a number of useful properties; for example, (i) they reduce the number of variables in the problem (e.g. three fitnesses reduced to two numbers, s and t; (ii) they express the *relation* between variables; (iii) the DLNs, being unit-free, have magnitudes that have absolute meaning from case to case ($s = 0.2$, $t = 0.5$ means the same thing with respect to genotype dynamics, independent of the actual survival times fertility values).

DLNs have sometimes been used to characterize life histories. Five examples will show this: (i) sex ratio (proportion males) in a brood (Charnov 1982); (ii) reproductive effort (Williams 1966) loosely defined as the proportion of available resources devoted to reproduction, as opposed to growth or maintenance; (iii) scaling allometry of life-history variables with body size (Harvey *et al.* 1989; Calder 1984; Millar & Zammuto 1983), the exponents are DLNs (indeed, deviations from log-scaling relations are also DLNs; Harvey *et al.* 1989); (iv) the dimensionless form of age-of-sex-change in a sequential hermaphrodite (Charnov & Bull 1989); and finally (v) the total force of mortality over some life-history phase, such as birth to adulthood (Ricklefs 1969; Charnov 1991). In each of these cases, empiricists have noted some general patterns (e.g. mortality rates scale with the -0.25 power of body size in mammals, or the first sex is always more abundant among the breeders under sex change) and life-history theorists have attempted general explanations as to why natural selection in the face of trade-offs has produced the patterns.

This paper reviews recent efforts to apply some other DLNs to characterize animal (and plant) life histories, in particular some numbers developed in the context of fishery science, which summarize relations between growth, mortality, and maturation (Beverton & Holt 1959). We review the Beverton–Holt patterns for fish, and extend the results to sea urchins, shrimp, snakes, and lizards. In the process we propose a new DLN related to theirs (Charnov & Berrigan 1990), and use this one to look additionally at birds and mammals. Finally, we show how evolutionary life-history theory may be used to answer why some of the patterns exist. Our claim is that the 'aggregate' characterization of life histories through these DLNs leads us to see new patterns in the data and to develop evolutionary life-history theory in novel ways.

2. LIFE HISTORIES WITH DETERMINATE AND INDETERMINATE GROWTH

Birds, mammals, insects (and a few other animals) have determinate growth where adult size does not alter. By contrast most other animals have inde-

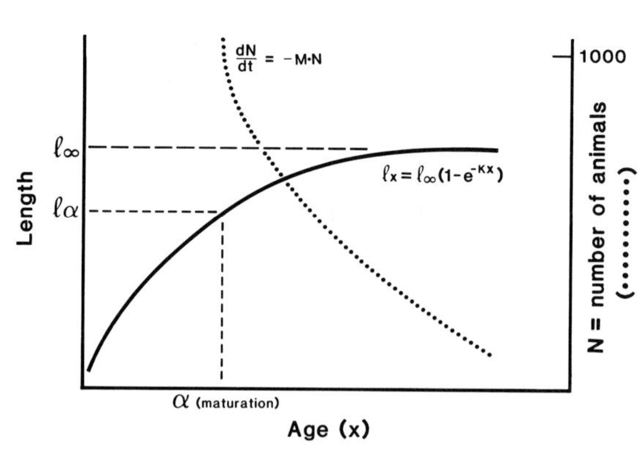

Figure 1. A life history under indeterminate growth. Growth in length (l_x) follows the Bertalanffy equation, $l_x = l_\infty(1-e^{-Kx})$, mass follows length to a power near 3. Maturation is at age α, length l_α, while l_∞ is the asymptotic length; thus l_α/l_∞ is the relative size at maturity. The adult instantaneous mortality rate is M. In practice M is a weighted average over the adult lifespan, weighted towards the younger adults (see Beverton (1963) for the specific statistic).

terminate growth where body size continues to increase after maturation. Figure 1 shows a typical life history. Body length (l) increases with age and is often usefully described by the Bertalanffy equation; in its simplest form the growth equation has two parameters, l_∞, the asymptotic length, and K, the growth coefficient. Maturation is at age α, corresponding to size l_α. In this simple example (figure 1) the adult instantaneous mortality rate (M) is shown as a constant, resulting in exponential decline for the cohort reaching the age of maturity. In actual practice M may increase over the adult life time; here we will be led to define an average M (Beverton 1963). K and M both have dimensions of 1/time, l_∞ and l_α have dimensions of length, and α has the dimension of time. From these we can construct three DLNs:

l_α/l_∞ = relative size at maturity (1)

$\alpha \cdot M$ = relation between maturation and adult mortality (ratio of age at maturity (α) to the average adult lifespan ($1/M$) (Charnov & Berrigan 1990)) (2)

K/M = relation between relative growth (K) and mortality (M). (3)

The number $K \cdot \alpha$ is related to l_α/l_∞ through the Bertalanffy equation of figure 1; or

$$\frac{l_\alpha}{l_\infty} = 1 - e^{-K \cdot \alpha}. \qquad (4)$$

Notice also that equation (4) implies that any two of DLNs (l_α/l_∞, K/M or $M \cdot \alpha$) suffice to determine the third.

The suggestion that we view life histories under indeterminate growth in terms of the two DLNs K/M and l_α/l_∞ goes back 30 years and is owing to Beverton & Holt (1959) and Beverton (1963), who developed the notion in relation to fish; they were motivated by the fact that in fisheries the steady-state equation for annual catch per recruit (a dimensionless equation) could be written in a form that included only three parameters, all of which were DLNs, two of which were K/M and l_α/l_∞. (Our use of M here differs slightly from theirs but the point here is simply to acknowledge their priority.) They were thus led to ask if various fish groups showed any patterns in the values of K/M and/or l_α/l_∞.

Determinate growers, like birds, mammals, and insects have l_α/l_∞ near 1; the $\alpha \cdot M$ number is useful to characterize them. There are, however, additional possibilities. With a clutch size per unit time per mother of b, independent of age, and survival to adulthood (age α) written as S, the average number of offspring produced over an individual's life is $b \cdot S/M$. Since b has units of 1/time, this equation is dimensionless and with a 1:1 secondary sex ratio will equal 2 in a non-growing population (Charnov 1986; Sutherland *et al.* 1986). We can rewrite it as $(\alpha \cdot b)/(\alpha \cdot M) S = 2$, which give a relation between three DLNs ($S, \alpha \cdot b, \alpha \cdot M$) imposed by the condition of population stability. This paper will only deal with the $\alpha \cdot M$ number for the determinate growers. Charnov (1991) discusses the other two for female mammals.

This paper will review some of the empirical patterns relating to the above DLNs, and will develop evolutionary life-history theory about the $\alpha \cdot M$ number. We begin with the classical Beverton–Holt patterns for indeterminate growers.

3. BEVERTON–HOLT: FISH

Thirty years ago Beverton & Holt (1959) and Beverton (1963) pioneered the comparative study of fish life histories by showing that within limited taxonomic boundaries (such as within the cod, salmon or herring family), there existed certain across species (or populations within a species) patterns in growth and mortality. These patterns, reviewed in Cushing (1968) and Pauly (1980) are two in number. (A third pattern, not developed here, is discussed in Charnov & Berrigan (1991).) Within each taxon the adult instantaneous mortality rate, M, and the Bertalanffy growth coefficient, K, are positively related to each other so that the ratio $K:M$ tends to be relatively constant; and the $K:M$ ratio differs between taxa. The second pattern is that the length at maturity (l_α) is positively related to the Bertalanffy asymptotic length (l_∞) so that the relative length at maturity, l_α/l_∞, tends to be a constant value within a taxon. Of course, as shown with equation (4), these two imply the constancy of $\alpha \cdot M$ within a taxon. Figure 2 shows an example of the data, here for the Clupeomorph fishes of the families Clupeidae and Engraulidae. Figure 2*a* shows a plot of $1/T_{max}$ versus K, where T_{max} is the age of the oldest individual observed in a large sample. Beverton (1963) showed that, at least for large samples, many fish species or populations have maximum age (T_{max}) that is highly correlated with the adult mortality rate M, so that $M = g/T_{max}$ with $g \simeq 6$. In a much larger and taxonomically diverse sample of animal species, Hoenig (1983) confirmed Beverton's relation, with a similar g value. Applied to figure 2, this relation has

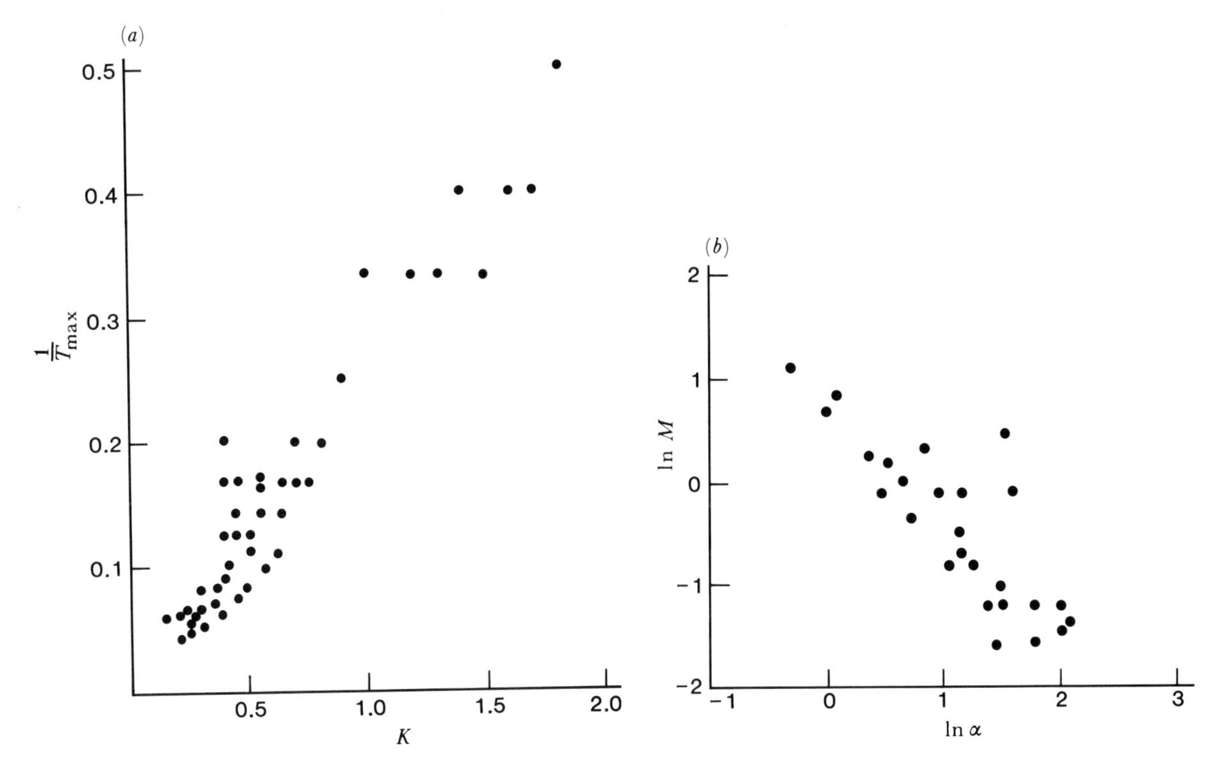

Figure 2. (*a*) Growth coefficient, K, against the inverse of the maximum lifespan ($1/T_{max}$) for 45 populations of ten species of fish in the families Engraulidae and Clupeidae. Because the adult instantaneous mortality rate, M, is proportional to $1/T_{max}$ (or $M \simeq 6/T_{max}$), we have that $M \simeq 1.5$ K for these fish. Figure redrawn from Beverton (1963). (*b*) The adult instantaneous mortality rate, M, is inversely proportional to the age of maturity, α, for fish in the families Clupeidae and Engraulidae, or $\ln M = C - \ln \alpha$. Data from Beverton (1963) and only includes populations with direct estimates of M (which is why there are fewer data points here compared with figure 2*a*). ($y = 0.69 - 1.04x$; $r = -0.84$; $n = 26$; s.t.d. errors; slope $= 0.14$; intercept $= 0.17$.) Time in years.

$M \simeq 1.5 \cdot K$ for the Clupeomorpha. Figure 2*b* shows that M and α are inversely proportional, making $M \cdot \alpha \simeq e^{0.69} = 2$. These two make $l_\alpha/l_\infty \simeq 0.75$, a number confirmed by the length data. While Beverton (1963) discusses some between-species differences in these numbers, the overall pattern is near constancy.

4. OTHER INDETERMINATE GROWERS

K/M, l_α/l_∞ and $\alpha \cdot M$ may also be approximately constant within other taxa showing indeterminate growth. Ebert (1975) showed K/M near 1 in a sample of over a dozen species of sea urchin ($r = 0.91$, sample size $= 15$, line through the origin). He provided no data on $\alpha \cdot M$ or l_α/l_∞. Charnov (1979, 1989) showed all the Beverton–Holt patterns to hold within the shrimp family Pandalidae in a sample that included 27 populations of five species and spanned the Northern latitudes from California to the subarctic. The data have $K/M \simeq 0.37$, $l_\alpha/l_\infty \simeq 0.56$, and $\alpha \cdot M \simeq 2.2$.

The above indeterminate growers are aquatic ectotherms. Shine & Charnov (1991) asked if the Beverton–Holt patterns also held for terrestrial ectotherms. They assembled data for 16 species of snakes and 20 species of lizards. l_α/l_∞ is near a constant for both snakes ($\simeq 0.64$) and lizards ($\simeq 0.73$). M and α are also inversely proportional in these groups, with $\alpha \cdot M \simeq 1.3$ for lizards and a bit higher ($\simeq 1.5$) for snakes.

It appears that the Beverton–Holt fish patterns (the approximate constancy of $\alpha \cdot M$, l_α/l_∞ and K/M 'within a kind-of-animal') hold for several other groups with indeterminate growth; *the life-history rules are about the allowed values for certain dimensionless numbers.* Now we turn to the determinate growers, birds and mammals.

5. $\alpha \cdot M$ IN BIRDS, MAMMALS (AND THE OTHERS)

Bird and mammals both have determinate growth but with one key difference. Mammals begin reproducing at near 90% their adult mass while birds with altricial young usually reach their adult mass near the time of independence from the parents. Most birds reach adult size long before α while mammals reach it near α.

They also differ in the $\alpha \cdot M$ number. Figure 3 shows a plot of $1/M$ (average adult lifespan) versus α for 66 bird species and 26 mammal species; the birds have adult lifetimes about double a mammal with the same age of maturity (recall that $1/M$ divided by α is $1/(\alpha \cdot M)$). $\alpha \cdot M$ is $\simeq 0.70$ for mammals, $\simeq 0.40$ for birds. A plot of $\ln M$ versus $\ln \alpha$ shows a slope of -1 for the mammals, but a somewhat steeper slope (-1.2) for the birds; thus $1/M$ is proportional to α in mammals but only approximately so for the birds. Birds with higher α^s have slightly higher $1/M^s$ than expected by strict proportionality (table 1).

Of course, the other way to summarize these relations is to plot the average length of the adult lifespan (expectation of further life at age α) ($\simeq 1/M$) versus the age at maturity (α). We summarize these

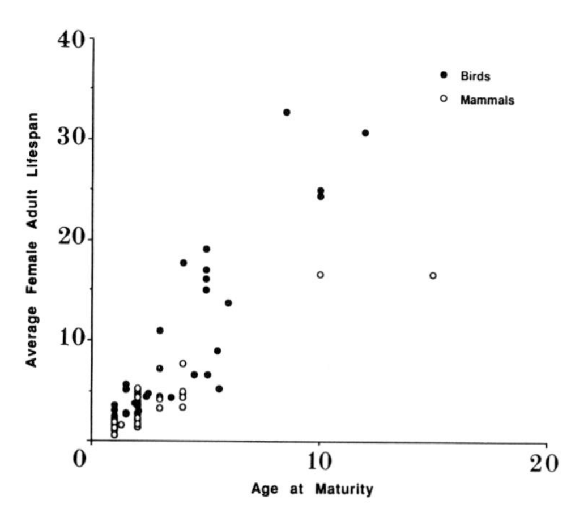

Figure 3. Average female adult lifespan ($1/M$) plotted against age of maturity (α) for 26 mammal species (data from Millar & Zammuto (1983) and Promislow & Harvey (1990)) and 66 bird species (data from many sources). Statistics as follows: mammals: $y = 0.27 + 1.25x$ ($r = 0.95$, $n = 26$); birds; $y = -1.32 + 2.75x$ ($r = 0.91$, $n = 66$). For a given α, birds have adult lifespans near double a comparable mammal. See also figure 4.

Table 1. *A summary of $\alpha \cdot M$ for all the groups*

group	$\alpha \cdot M$
birds	$\simeq 0.40$
mammals	$\simeq 0.70$
snakes and lizards	$\simeq 1.40$
fish and shrimp	$\simeq 2.00$

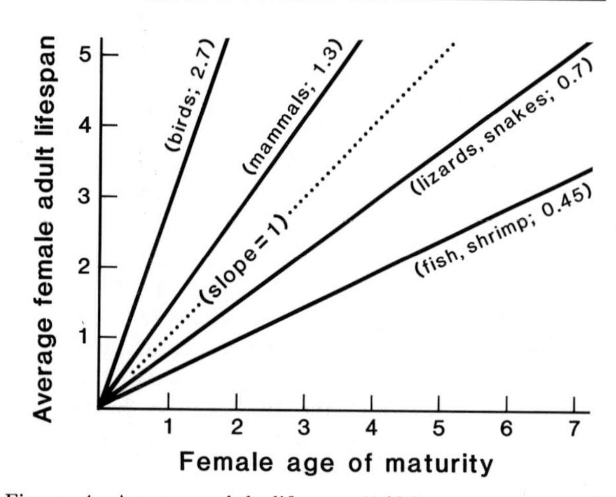

Figure 4. Average adult lifespan ($1/M$) versus α for all groups. For contrast I treat the birds as having a proportional relation (see text for qualifications). I also pool lizards and snakes. Shrimp data are for both sexes. Number on each line refers to estimated slope, the number $1/(\alpha \cdot M)$. (After Charnov & Berrigan 1990.)

data in figure 4. Notice that birds have adult lifetimes, for a given age of maturity, about 2 × mammals, 3 × snakes and lizards, and 5 × fish and shrimp. Notice also that figure 4 says nothing about the actual values of α or M; indeed, there are fish (Beverton 1963; Beverton & Holt 1959) with ages of maturation of 10–20 years, the same magnitude as elephants and well beyond most birds. It is the *relation* between α and M which differs between groups; upon reaching the maturation age of 10 years, a bird has an average of about 25 years to live, a fish has 5 years. These differences are so striking that they demand explanation in terms of some fundamental differences between the groups (Charnov & Berrigan 1990).

6. WHAT SETS $\alpha \cdot M$?

The above heading could well expand to include K/M and l_α/l_∞, but this paper will only deal with $\alpha \cdot M$. Charnov & Berrigan (1991) have begun a life-history evolution theory aimed at predicting all three DLNs under indeterminate growth. Our aims are more modest here ($\alpha \cdot M$ alone) and we ignore the added complications of indeterminate growth.

We believe that the answer to why $\alpha \cdot M$ takes on particular values lies in how natural selection acts to set the maturation time itself. In what follows we will construct a theory for evolution of the age of maturity and require as its output the $\alpha \cdot M$ number. The broad brush approach to the theory will be in three phases: first, a general evolutionary theory for α; second, a phenomenological approach to predicting the $\alpha \cdot M$ number; and finally an 'individual growth or productivity' approach specifically designed for determinate growers like mammals.

The model

Consider a newborn female and define l_x as the probability she is alive at age x, and b_x as her birth rate, in daughters, at age x. Her lifetime production of daughters is:

$$R_o = \int_\alpha^\infty l_x b_x \, dx. \tag{5}$$

We can rewrite R_o as follows:

$$R_o = l_\alpha \left[\frac{\int_\alpha^\infty l_x b_x \, dx}{l_\alpha} \right]. \tag{6}$$

The term in brackets is the average number of daughters born over a female's adult lifespan, the 'reproductive value' (Fisher 1930) of an age α (a just mature) female, and will therefore be labelled $V(\alpha)$.

Now, write l_α as $e^{-\phi(\alpha)}$. R_o can now be written as:

$$R_o = e^{-\phi(\alpha)} \cdot V(\alpha). \tag{7}$$

For R_o to be a valid fitness measure, the population must not be growing, or $R_o \approx 1$. This is a population dynamic side condition on an optimization-of-R_o problem (Charnov 1986) and is discussed in great detail in Charnov (1990). We wish to maximize R_o with respect to α, which is the same as maximizing:

$$\ln R_o = \ln V(\alpha) - \phi(\alpha).$$

In equilibrium (at the ESS α, Maynard Smith (1982)), we require

$$\frac{\partial \ln V(\alpha)}{\partial \alpha} = \frac{\partial \phi(\alpha)}{\partial \alpha}. \tag{8}$$

Now, suppose that $Z(x)$ is the instantaneous mortality rate at age x; in general $Z(x)$ will decrease with x (and for many species will reach some low and near constant value before maturation; it may go up again late in life). We may thus write

$$\phi(\alpha) = \int_0^\alpha Z(x)\,\mathrm{d}x$$

and

$$\frac{\mathrm{d}\phi(\alpha)}{\partial\alpha} = Z(\alpha).$$

But if mortality does not change much after maturation $Z(\alpha)$ is the adult mortality rate, called M earlier in this paper. The ESS equation (8) may now be written as

$$\frac{\partial \ln V(\alpha)}{\partial\alpha} = M. \qquad (9)$$

This equation for the ESS age of maturity is the first step towards getting a value for the $\alpha \cdot M$ number. The key apparently lies in the $V(\alpha)$ function; notice that equation (9) does not require that we know the actual $V(\alpha)$ function, only its proportional change with α, its shape (Charnov 1990). For example, suppose that we guess that $V(\alpha) \propto \alpha^d$; $V(\alpha)$ is a power function in α with exponent (a DLN!) d. $\ln V(\alpha) = \text{constant} + d \ln \alpha$, and

$$\frac{\partial \ln V(\alpha)}{\partial\alpha} = d/\alpha.$$

If we put this into equation (9), a rather interesting thing happens; the ESS is where $\alpha \cdot M = d$. All life histories where $V(\alpha)$ can be treated as a power function in α have the property that '$\alpha \cdot M$ = the exponent' at the ESS. Here is a theory for the $\alpha \cdot M$ number; it suggests that fish have quite high exponents and that birds have quite small ones. Better still, we know the values of d (at least approximately) to be searched for. This is a phenomenological model as nothing really informs us as to what determines the d coefficient, only that whatever it is is similar within fish, birds, etc. It seems clear that to go further we must tie d (or something like it) back to general models of growth, or other developmental processes.

7. $\alpha \cdot M$ IN MAMMALS

In this section we model $V(\alpha)$ as a function of individual productivity; the approach makes two new assumptions, in addition to those leading to equation (9). These are that growth depends on body size (W) and can be described as $\mathrm{d}W/\mathrm{d}T = AW^c$ (equation 10), and that growth is determinate and ceases at reproductive maturity when energy is simply diverted from growth to offspring production. The derivation in Appendix 1 shows that these assumptions and equation (9) lead to the prediction that $\alpha \cdot M = (c/1-c)\,(1-\delta^{1-c})$ (equation 11), where c is the exponent describing the size dependence of energy acquisition for growth and reproduction (equation 10) and δ is the offspring's relative size at independence, its mass at independence divided by its mother's mass. Direct measurements of animal production rates put c near 0.75 within many taxa (Lavigne 1982). Note that if $c = 0.75$, equation (11) reduces to $3(1-\delta^{0.25})$ (equation 12). The relative size at independence (δ) is the point at which an animal moves from a potential growth trajectory determined by its primary care giver to one determined by its own size. Thus we are assuming that a mammal grows during two periods; one from birth to independence where its mother controls its growth rate and a second from independence to maturity, where its own size determines its growth rate. The size at maturity is assumed to be the final adult size and this size in turn determines the offspring production rate through equation (10). The age at maturity (α) is the interval between independence (estimated here as weaning) and first reproduction.

Equation (12) is almost linear in δ for $0.2 < \delta < 0.6$, a range that includes most mammals (Millar 1977). This means that to predict the average $\alpha \cdot M$, it is sufficient to substitute the average δ ($= \bar{\delta}$) for mammals. Millar (1977) estimated $\bar{\delta}$ (weaning mass/adult female mass) for 100 species, mostly < 1 kg in mass, and got $\bar{\delta} = 0.37$. We have an additional sample of 23 species (Appendix 2), mostly of body size > 1 kg and get $\bar{\delta} = 0.33$. These numbers inserted into equation (12) predict $\alpha \cdot M$ to be ≈ 0.7, right at the observed average value.

We have also tested the prediction that $\alpha \cdot M = 3(1-\delta^{0.25})$ by comparing the values of $\alpha \cdot M$ with the ratio of mass at weaning to the average adult female mass (δ) for 23 species of mammals (figure 5; data sources and species listed in Appendix 2). The data strongly support the predicted relation. Notice that the r value of the linear regression between observed and predicted values of $\alpha \cdot M$ increases from 0.71 to 0.93 when we fit averages (figure 5*b*) over even intervals of δ rather than all 23 points (figure 5*a*). We cannot distinguish between a linear regression of $\alpha \cdot M$ versus δ ($r = -0.67$, $p < 0.001$) and this slightly curvilinear relation ($r = 0.71$, $p < 0.001$). In these tests we assume that $c = 0.75$. We also used nonlinear regression to fit the one parameter model for c (equation 11). The solution from fitting all 23 points, plus or minus one standard error, gives a value of $c = 0.74 \pm 0.03$ ($r = -0.71$, $p < 0.001$) and the solution for the five average values gives $c = 0.75 \pm 0.04$ ($r = -0.93$, $p < 0.03$). Because the production relation of equation (10) appears to be an important component of mammalian life histories, it is particularly significant that at evolutionary equilibrium, the 0.75 exponent appears in the relation between $\alpha \cdot M$ and δ (equation 12). Our analysis (figure 5) is the first indirect determination of this exponent and gives virtually the same answer as direct measurements of individual or offspring production (Lavigne 1982).

The ecological correlates of differences in $\alpha \cdot M$ and δ within the mammals are not obvious. The three largest values of δ in this study are those of the impala, wildebeest, and zebra, and the three lowest those of the rabbit, otter, and boar; squirrels and elephants have similar and intermediate values of δ and $\alpha \cdot M$. Body size is also not responsible for the observed correlation between $\alpha \cdot M$ and δ because $\alpha \cdot M$ is not correlated with adult mass ($r < 0.001$, $p > 0.05$) or mass at

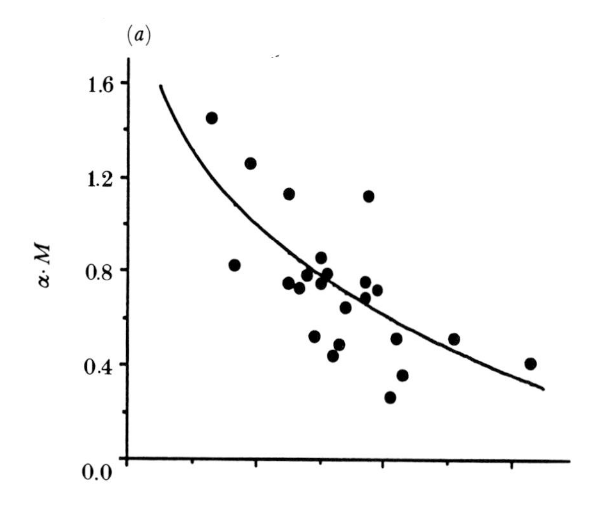

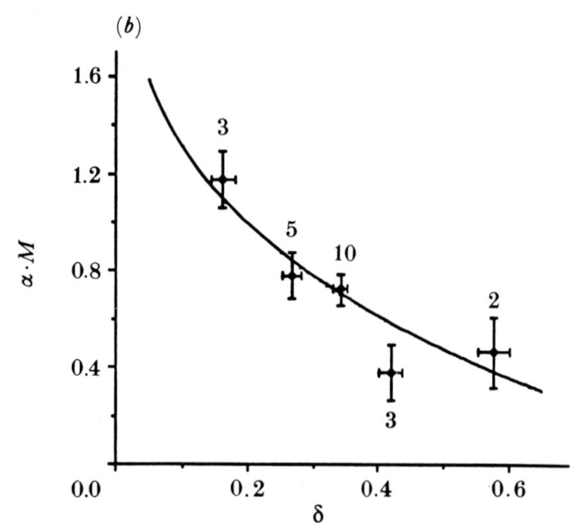

Figure 5. We plot the value of $\alpha \cdot M$ versus δ for females of 23 mammal species (see Appendix 2). α is the age at first reproduction adjusted for the parental care period. M is the average adult instantaneous mortality rate (Beverton 1963) and δ is the ratio of mass at weaning to adult female mass. (*a*) We show the observed values of $\alpha \cdot M$ (●) and the predicted values (equation 12) (—). We tested the fit with a linear regression of observed against predicted values ($r = 0.71$, $p < 0.001$). (*b*) Here we averaged the values of δ and $\alpha \cdot M$ over the intervals $\delta = 0.1$–0.2, 0.2–0.3, 0.3–0.4, 0.4–0.5, and > 0.5. The error bars are one standard error, the numbers are the sample size, and the line is the predicted value ($r = 0.93$, $p < 0.03$). The statistical tests are one-tailed.

weaning ($r < 0.003$, $p > 0.05$). Considering the uncertainties involved in field estimates of life-table parameters and the estimation of δ by weaning mass, the observed fit of the data to the predicted line is encouraging.

8. DISCUSSION AND CONCLUSION

Perhaps the most appealing aspect of a dimensionless approach to life histories is its focus on general patterns, in terms of relations between vital rates (growth, mortality, maturity, fecundity); and the data do show patterns (even for some plant groups, see, for example, Loehle (1988) and figure 2 therein), begging for explanation in terms of evolutionary life-history theory (which itself must be made dimensionless). The exploration of these patterns, in both theory and data, has really just begun. Life-history theory is almost always based on the assumption of fitness maximization in the face of trade-offs (= constraints). If the fitness function (e.g. R_o) does not differ between species, then species differ in life history only because they differ in trade-offs. However, similarities of DLNs (within a taxon) strongly suggest similarities in the trade-off functions. Indeed, several modelling efforts (Charnov 1990, 1991; Charnov & Berrigan 1991) say that similarities of the DLNs point to similarities of the shapes of the trade-offs (e.g. $\alpha \cdot M \approx 1$ implies $d \approx 1$, under the assumption $V(\alpha) \propto \alpha^d$). Another way to say this is that the various DLNs show *conservation principles* (i.e. they are fairly invariant within a taxon) and that the existence of such conserved quantities points to deeper underlying symmetries (Watson 1990) in the transitions allowed for trade-offs; the allowed trade-offs may differ in height but not in shape.

We would like to conclude this paper with a brief and final illustration of the above points, chosen from the theory of sex change. Sex change exists in two forms: protandry, male first and protogyny, female first. The usual evolutionary theory aims to predict the order and time of sex change (Charnov 1982). The dimensionless version of sex-change theory aims to predict the breeding sex ratio. Under quite broad conditions, the theory says that the first sex will be more abundant among the breeders, and that the extent of bias depends upon just how fast (= shape) each sex gains reproductive ability with age or size (Charnov 1982; Charnov & Bull 1989). Indeed, the first sex is almost always more abundant, and the skew is greater under protogyny (Charnov & Bull 1989). The dimensionless view of sex change provides us with some broad empirical rules, and some general theoretical reasons (hypotheses) as to why the data patterns exist.

We have benefited greatly from discussions with Raymond Beverton, Jan Kozlowski, Robert Ricklefs, Paul Harvey, Sean Nee, Linda Partridge and Dave Stephens. We thank K. Ralls, E. Rickart and T. Nagy for help with data collection.

REFERENCES

Beverton, R. J. H. 1963 Maturation, growth and mortality of clupeid and engraulid stocks in relation to fishing. *Rapp. Proces-Verb. Cons. Intern. Explor. Mer.* **154**, 44–67.

Beverton, R. J. H. & Holt, S. J. 1959 A review of the lifespans and mortality rates of fish in nature and the relation to growth and other physiological characteristics. In *Ciba Foundation, Colloquia in ageing. V. The lifespan of animals*, pp. 142–177. London: Churchill.

Calder, W. A. 1984 *Size, function and life history.* Harvard University Press.

Charnov, E. L. 1979 Natural selection and sex change in Pandalid shrimp: test of a life history theory. *Am. Nat.* **113**, 715–734.

Charnov, E. L. 1982 *The theory of sex allocation.* Princeton University Press.

Charnov, E. L. 1986 Life history evolution in a 'recruitment population': why are adult mortality rates constant *Oikos* **47**, 129–134.

Charnov, E. L. 1989 Natural selection on age of maturity in shrimp. *Evol. Ecol.* **3**, 236–239.

Charnov, E. L. 1990 On evolution of age of maturity and the adult lifespan. *J. evol. Biol.* **3**, 139–144.
Charnov, E. L. 1991 Evolution of life history variation among female mammals. *Proc. natn. Acad. Sci. U.S.A.* **88**, 1134–1137.
Charnov, E. L. & Bull, J. J. 1989 Non-fisherian sex ratios with sex change and environmental sex determination. *Nature, Lond.* **338**, 148–150.
Charnov, E. L. & Berrigan, D. 1990 Dimensionless numbers and life history evolution: age of maturity versus the adult lifespan. *Evol. Ecol.* **4**, 273–275.
Charnov, E. L. & Berrigan, D. 1991 Evolution of life history parameters in animals with indeterminate growth, particularly fish. *Evol. Ecol.* **5**, 63–68.
Cushing, D. H. 1968 *Fisheries biology*. University Wisconsin Press.
Ebert, T. 1975 Growth and mortality in post-larval echinoids. *Am. Zool.* **15**, 755–775.
Fisher, R. A. 1930 *The genetical theory of natural selection.* Oxford University Press.
Giordano, F. R., Wells, M. E. & Wilde, C. O. 1987 Dimensional analysis. In *COMAP: tools for teaching 1987*, pp. 72–97. Arlington, MA: COMAP.
Harvey, P. H., Read, A. F. & Promislow, D. E. L. 1989 Life history variation in placental mammals: unifying the data with theory. *Oxf. Surv. evol. Biol.* **6**, 13–31.
Hoenig, J. M. 1983 Empirical use of longevity data to estimate mortality rates. *Fish. Bull.* **82**, 898–903.
Kozlowski, J. & Wiegert, R. G. 1986 Optimal allocation of energy to growth and reproduction. *Theor. Popul. Biol.* **29**, 16–37.
Kozlowski, J. & Wiegert, R. G. 1987 Optimal age and size at maturity in annuals and perennials with determinate growth. *Evol. Ecol.* **1**, 231–244.
Lavigne, D. M. 1982 Similarity of energy budgets of animal populations. *J. Anim. Ecology* **51**, 195–206.
Loehle, C. 1988 Tree life histories: the role of defenses. *Can. J. For. Res.* **18**, 209–222.
Maynard Smith, J. 1982 Evolution and the theory of games. Cambridge University Press.
Millar, J. S. 1977 Adaptive features of mammalian reproduction. *Evolution* **31**, 370–386.
Millar, J. S. & Zammuto, R. M. 1983 Life histories of mammals: an analysis of life tables. *Ecology* **64**, 631–635.
Pauly, D. 1980 On the interrelationships between natural mortality, growth parameters, and mean environmental temperature in 175 fish stocks. *J. Cons. Cons. Int. Explor. Mer.* **39**, 175–192.
Promislow, D. E. L. & Harvey, P. H. 1990 Living fast and dying young: a comparative analysis of life history variation among mammals. *J. Zool.* **220**, 417–437.
Reiss, M. J. 1989 *The allometry of growth and reproduction.* Cambridge University Press.
Ricklefs, R. E. 1969 Natural selection and the development of mortality rates in young birds. *Nature, Lond.* **223**, 922–925.
Shine, R. & Charnov, E. L. 1991 Patterns of survivorship, growth and maturation in snakes and lizards. *Am. Nat.* (Submitted.)
Stahl, W. R. 1962 Similarity and dimensional methods in biology. *Science, Wash.* **137**, 205–212.
Stephens, D. W. 1991 Dimensional analysis in behavioral ecology. (MS.)
Sutherland, W. J., Grafen, A. & Harvey, P. H. 1986 Life history correlations and demography. *Nature, Lond.* **320**, 88.
Watson, A. 1990 The mathematics of symmetry. *New Scient.* **1740**, 45–50.
Williams, G. C. 1966 Natural selection, the cost of reproduction and a refinement of Lack's principle. *Am. Nat.* **100**, 687–690.

APPENDIX 1 $\alpha \cdot M$ FOR MAMMALS

The growth and production model developed here is strictly applicable only to determinate growers like female mammals who cease growing at age α. The argument is from Kozlowski & Wiegert (1986, 1987), and Charnov (1990, 1991).

Let W stand for body mass; then a great many groups of animals show the following growth relation (before maturation but after independence from parental feeding):

$$\frac{dW}{dT} = AW^c, \tag{1}$$

where c is near 0.75 and A varies between taxonomic groups; see Lavigne (1982) and review by Reiss (1989). Set $c = 0.75$; then we have

$$\int_{W_0}^{W(\alpha)} \frac{dW}{W^{0.75}} = \int_0^{\alpha} AT, \tag{2}$$

where W_0 is the size at the end of parental feeding, called time zero. The change of variable $Y = W^{0.25}$ leads to $(dY = [0.25]/[W^{0.75}]\, dW)$ and the general solution of equation (2):

$$W(\alpha)^{0.25} - W_0^{0.25} = \frac{A\alpha}{4}.$$

Now, write W_0 as $\delta \cdot W(\alpha)$; we have finally

$$W(\alpha)^{0.25} = \frac{0.25 \cdot A}{1-\delta^{0.25}} \cdot \alpha \quad (\delta \text{ is of course a DLN}). \tag{3}$$

Notice that α in equation (3) is measured from some time called zero when we assign the individual some starting size W_0 which is taken to be δ proportion of the adult size.

Equation (3) is simply a growth relation. Kozlowski & Wiegert (1986, 1987) noted that equation (1) is also an offspring production relation if offspring are simply the result of shifting resources, primarily energy, from self-growth to offspring-production. Let b = offspring production (per female) per unit time, then for a determinate grower $V(\alpha) = b/M$ (see equation (7) in text for $V(\alpha)$). Provided M does not increase with a delay in maturation (the mortality rate reaches its minimum value prior to age α) then $V(\alpha) \propto b$. But if $b \propto dW/dT \propto AW^{0.75}$, then $V(\alpha) \propto AW^{0.75}$, $\ln V(\alpha) =$ Constant $+ 0.75 \log W$ and

$$\frac{\partial \ln V(\alpha)}{\partial \alpha} = \frac{0.75\, dW}{W\ dT} = 0.75\ AW^{-0.25}.$$

From equation (9) in the text we have the ESS result

$$M = \frac{\partial \ln V(\alpha)}{\partial \alpha} = 0.75AW^{-0.25}. \tag{4}$$

If we use the growth equation (3) to eliminate W from (4), the following results:

$$\alpha \cdot M = 3(1-\delta^{0.25}). \tag{5}$$

For an arbitrary exponent c in equation (1), we have in general

$$\alpha \cdot M = \frac{c}{1-c}[1-\delta^{1-c}]. \tag{6}$$

Notice that this argument (equations 3 and 4) also gives the known ± 0.25 scaling of age at maturity (α) and mortality (M) with adult body size (Harvey *et al.* 1989).

APPENDIX 2

Table 1. *Data on $\alpha \cdot M$, and δ for 23 species of mammals* (The adult mortality rate (M) and age at first reproduction (α) (from estimated age at independence) were obtained from Millar & Zammuto (1983). The ratio of mass at weaning to adult mass (δ) was obtained from the sources listed below.)

species	$\alpha \cdot M$	δ	source for δ
Castor canadensis	1.13	0.25	Aleksiuk & Cowan (1969)
Sciurus carolinensis	0.52	0.29	Horwich (1972)
Spermophilus armatus[a]	0.78	0.28	Slade & Balph (1974)
S. beldingi	0.75	0.30	Morton & Tung (1971)
S. lateralis	0.76	0.37	Millar (1977)
S. parryi	0.79	0.31	Armitage (1981)
Tamias striatus	0.86	0.30	Wishner (1982)
Tamiascurus hudsonicus	0.44	0.32	Millar (1977)
Ochotona princeps	0.52	0.42	Millar (1977)
Sylvilagus floridanus	1.45	0.13	Millar (1977)
Lutra canadensis	0.82	0.17	Mason & Macdonald (1986)
Lynx rufus	0.73	0.27	Crowe (1975)
Mephitis mephitis	0.65	0.34	Casey & Webster (1975)
Taxidea taxus	0.49	0.33	Neal (1986)
Equus burchelli	0.42	0.63	Wackernagel (1965)
Aepyceros melampus	0.36	0.43	Howells & Hanks (1975)
Cervus elaphus	1.13	0.37	Clutton-Brock *et al.* (1982)
Connochaetes taurinus	0.52	0.51	Talbot & Talbot (1963)
Kobus defassa	0.27	0.41	Spinage (1982)
Ovis canadensis	0.72	0.39	Hansen & Deming (1980)
Sus scrofa	1.26	0.19	Myrcha & Jezierski (1972)
Syncerus caffer	0.69	0.37	Sinclair (1977)
Loxodonta africana	0.75	0.25	Laws (1966)

[a] Averaging the four values for the *Spermophilus* spp did not significantly affect the results of the analysis shown in the text.

REFERENCES FOR APPENDIX 2

Aleksiuk, M. & Cowan, I. M. 1969 The winter metabolic depression in arctic beavers (*Castor canadensis* Kuhl) with comparisons to California beavers. *Can. J. Zool.* **47**, 965–979.

Armitage, K. B. 1981 Sociality as a life-history tactic of ground squirrels. *Oecologia* **48**, 36–49.

Casey, G. A. & Webster, W. A. 1975 Age and sex determination of striped skunks (*Mephitis mephitis*) from Ontario, Manitoba, and Quebec. *Can. J. Zool.* **53**, 223–226.

Clutton-Brock, T. H., Guinnes, F. E. & Albon, S. D. 1982 *Red deer: behavior and ecology of two sexes.* University of Chicago Press.

Crowe, D. M. 1975 Aspects of ageing, growth, and reproduction of bobcats from Wyoming. *J. Mammal.* **56**, 177–198.

Hansen, C. G. & Deming, O. V. 1980 In *The desert bighorn* (ed. G. Monson & L. Sumner). The University of Arizona Press.

Horwich, R. H. 1972 The ontogeny of social behavior in the gray squirrel (*Sciurus carolinensis*). *Adv. Ethol.* **8**, 1–103.

Howells, W. W. & Hanks, J. 1975 Body growth of the Impala (*Aepyceros melampus*) in Wankie National Park, Rhodesia. *J. sth. Afr. Wildl. Mgmt. Ass.* **5**, 95–98.

Laws, R. M. 1966 Age criteria for the African elephant, *Loxodonta a. africana. East Afr. Wildl. J.* **4**, 1–37.

Mason, C. F. & Macdonald, S. M. 1986 *Otters.* Cambridge University Press.

Millar, J. S. 1977 Adaptive features of mammalian reproduction. *Evolution* **31**, 370–386.

Millar, J. S. & Zammuto, R. M. 1983 Life histories of mammals: an analysis of life tables. *Ecology* **64**, 631–635.

Morton, M. L. & Tung, H. L. 1971 Growth and development in the Belding ground squirrel (*Spermophilus beldingi beldingi*). *J. Mammal.* **52**, 611–616.

Myrcha, A. & Jezierski, W. 1972 Metabolic rate during postnatal development of Wild Boars. *Acta Therio.* **33**, 443–452.

Neal, E. 1986 *The natural history of badgers.* New York: Facts on File Publications.

Sinclair, A. R. E. 1977 *The African buffalo.* University of Chicago Press.

Slade, N. A. & Balph, D. F. 1974 Population ecology of Uinta ground squirrels. *Ecology* **55**, 989–1003.

Spinage, C. A. 1982 *A territorial antelope: the Uganda waterbuck.* London: Academic Press.

Talbot, L. M. & Talbot, M. H. 1963 The Wildebeest in Western Masailand. *East Africa. Wildl. Monogr.* no. 12.

Wackernagel, H. 1965 Grants' Zebra, *Equus burchelli boehmi*, at Basle zoo – a contribution to breeding biology. *Int. Zoo. Yearbook* **5**, 38–41.

Wishner, L. 1982 *Eastern Chipmunks.* Washington: Smithsonian Institute Press.

Discussion

R. J. H. Beverton (*Montana, Old Roman Road, Langstone, Gwent NP62JU, U.K.*). It is intriguing to see how Professor Charnov has been able to extend the early explorations by Holt and myself into dimensionless indices of life-history characteristics to a wide range of animal groups. Perhaps I could add two comments as a postscript. One concerns Professor Stearn's question (not printed) about the reality of the results. It is true that there is strong covariation in certain of the underlying parameters and it would certainly be unwise, for example, to attempt to apply detailed statistical tests of linearity to some of the relations. Nevertheless, the broad patterns of the basic ratios which Professor Charnov is comparing – such as that between age at maturity and longevity – are not artefacts; and his finding that they take clearly different characteristic values for the major animal groups – fish, reptiles, birds and mammals – opens up further rewarding avenues of study.

My other comment concerns the variation of these ratios within the environmental range of one species. The North American fish *Stizostedion vitreum* (walleye) provides a good example. At the southern end of its range, in Texas and Colorado, it matures (with difficulty) at 2 years and none live longer than 4 years, whereas in northern Canada it does not mature until it is about 7 or 8 years and lives to about 20; but the *size* at maturity and the total lifetime fecundity per maturing recruit is nearly the same throughout the environmental range. It is as if temperature is determining the 'rate of living' – manifest both in the time it takes to reach a threshold size at maturity and the subsequent lifespan, with growth and fecundity adjusted to achieve nearly the same overall 'fitness'. As John Thorpe pointed out at this meeting, there is much still to be learned about the physiological basis of the attainment of maturity.

An experimental exploration of Waddington's epigenetic landscape

LEO W. BUSS[1,2] AND NEIL W. BLACKSTONE[1]

Departments of Biology[1] and Geology and Geophysics[2], Yale University, New Haven, Connecticut 06511, U.S.A.

SUMMARY

Variation in the branching morphologies of clonal plants, fungi, and sessile marine invertebrates is frequently correlated with a suite of life-history traits (e.g. 'phalanx' or 'guerilla'). These correlations have been interpreted to be the causal product of selection. A tacit assumption of selection on a trait is that development is canalized in the manner Waddington originally suggested for aclonal taxa, i.e. small perturbations in development result in a return to an equilibrial morphology. We tested this assumption by manipulating developing colonies of the hydroid *Hydractinia echinata*. The growth trajectory of these colonies follows a clone-specific schedule of production of three structures: feeding polyps, stolonal mat, and peripheral stolons. Isogeneic manipulations of the relative frequency of these structures show that developing colonies can regulate the rate of production of these three structures, but that regulation does not result in rapid convergence on a common growth trajectory.

1. INTRODUCTION

Clonal life cycles are known in plants, animals and fungi and are widely appreciated as displaying patterns in ecology and evolution which differ profoundly from those characterizing aclonal taxa (recent symposium volumes include: Boardman *et al.* 1973; Larwood & Rosen 1979; Jackson *et al.* 1985; Harper 1986; Harper *et al.* 1986; and books include: Harper 1977; Buss 1987; McKinney & Jackson 1989; Caswell 1989). Although differences in demography and life history between clonal and aclonal taxa are pronounced, the substantive intellectual issues with which this symposium is grappling are identical in the two groups; only the life-history traits upon which we focus our attention differ. Yet clonal taxa offer some unique experimental options, precluded in aclonal taxa, which we hope to show are useful in testing tacit assumptions underlying current practice in the field of life-history theory.

Many clonal taxa are substrate-bound, encrusting organisms which advance over a surface by stolons, runners, rhizomes or analogous structures (Harper 1977; Jackson 1979). The vast majority of the fungi are of this habit as are colonial invertebrates in the phyla Porifera, Cnidaria, Bryozoa, as well as certain ascidians, annelids, entoprocts, phoronids, numerous algal groups, virtually all cryptogamic plants, and most herbaceous angiosperms. These groups constitute most of what covers the face of the planet (as opposed to what moves over the surface). In many groups, the morphology can be idealized as being composed of feeding and reproductive entities called ramets (e.g. zooid, polyp, plantlet) which are connected to other ramets, with varying degrees of permanence, by a vascular system (e.g. hyphae, rhizomes, stolons, funiculi). For simplicity, we will refer to all vascular connections as stolons. The spatial arrangement of ramets and stolons, in particular their modes of elongation and branching define a central life-history character of clonal taxa without obvious parallel in aclonal groups. It is on the interpretation of this character that we will focus our attention.

The branching patterns of clonal organisms have been variously interpreted as analogous to foraging behaviour, where the pattern of branching serves to locate patchily distributed resources, or to refuge seeking, where the pattern of branching defines a search strategy for the location of 'safe sites,' or as a competitive mechanism, wherein a pattern of branching reflects differential commitment to defending a particular site (Buss 1979; Jackson 1979; Lovett-Doust 1981; Bell 1984; Harper 1985; Salzmann 1985; Schmid 1985, 1986; Slade & Hutchings 1987; Sutherland 1987; Sutherland & Stillman 1988). Although an exhaustive review of this literature is beyond the scope of this study, an example of this reasoning is perhaps useful. Animal and plant biologists have independently contrasted groups that grow by the production of numerous, closely packed ramets with short vascular connections with groups that produce few, widely spaced ramets on long vascular connections (figure 1). These two ends of a spectrum have been shown to correlate with a variety of other life history characters (table 1).

The reasoning here is that which characterizes most life-history theorizing. One recognizes a life-history trait believed to be shaped by selection and reasons that trade-offs between that trait and other traits may define a suite of characters which can, in principle,

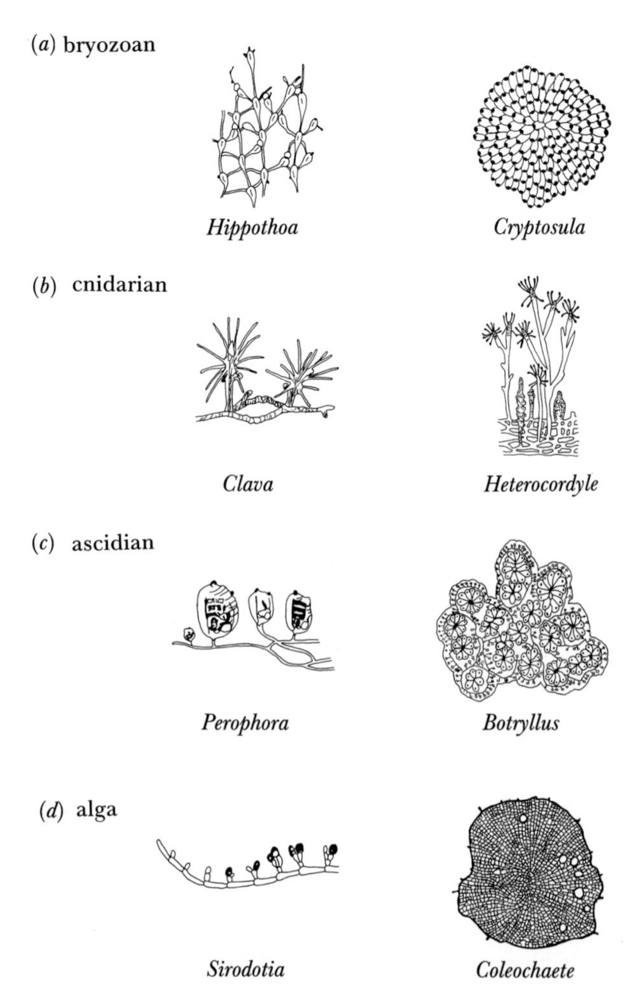

Figure 1. Examples of runner-like and sheet-like forms in (*a*) bryozoans, (*b*) cnidarians, (*c*) ascidians and (*d*) algae. Comparable variation is known in various fungal groups, in vascular plants, in several protist groups, and in sponges.

Table 1. *Comparison of life-history strategies of guerilla and phalanx forms*

life-history trait	phalanx compared with guerilla
size at first reproduction	larger
growth rate	slower
fecundity	lower
capacity for dispersal	lower
commitment to site of recruitment	higher
regenerative potential	higher
intra-colony reallocation of resources	higher
zooid polymorphism	higher
competitive ability	higher
morphological stability	higher

[a] drawn from tables in Buss (1979), Jackson (1979) and Harper (1985).

withstand interactions between it and an environment populated by organisms displaying a different suite of characters (Stearns 1976, 1977). Central to this reasoning is the concept of natural selection shaping an organism's response to trade-offs. While the power of this concept is undoubted, the problems with testing it are manifest: realistic evolutionary experiments can only be done with microbes, quantitative genetic signatures consistent with the operation of natural selection have alternative explanations, and correlations within and between taxa, however repeated, do not imply causality.

We here note that aspects of a selection-based interpretation of clonal organism morphology and life history are based on certain tacit assumptions with regard to the nature of development. Waddington represented development as a problem in dynamic systems theory (i.e. 'the epigenetic landscape'). His 'main thesis is that developmental reactions, *as they occur in organisms submitted to natural selection*, are in general canalized. That is to say, they are adjusted so as to bring about one definite end-result regardless of minor variations in conditions during the course of development' (Waddington 1942, p. 563, emphasis in the original). Thus, small disturbances to a developing organism should result in a return to an equilibrial morphology. To the extent that this conceptualization of the problem is a proper one, one might use adherence to Waddington's predictions as a proxy test for the role of selection in shaping life-history characters.

Let us restate the issue. The growth morphology of a clonal organism is known to be correlated with a suite of other life-history traits (table 1) and it is frequently claimed that this complex of traits is co-adapted. Because perturbations, in the forms of death of ramets, accidental severing of stolons, partial predation, and the like, are inevitable in natural populations, we would expect that a developing clone would respond to perturbations by return to its original morphology. That is to say, if selection is indeed favouring a particular morphology, then selection should act to buffer the development of that organism from those perturbations likely to be encountered (Waddington 1942). If we find that morphology is not canalized, then either our conceptualization regarding the relation between selection and canalization is flawed, or our hypothesis that selection underlies the correlation between life history-traits bears re-examination.

In clonal organisms, where development and growth are inseparable, Waddington's prediction can be directly subjected to an experimental test. We have performed experimental perturbations of developing colonies of the hydroid *Hydractinia echinata*. The growth trajectory of this species follows a clone-specific schedule of production of three structures: feeding polyps (ramets), stolonal mat, and peripheral stolons. Isogeneic manipulations of the relative frequency of these structures show that developing colonies can regulate the rate of production of these three structures, but that regulation does not result in rapid convergence on a common growth trajectory.

2. THE *Hydractinia* SYSTEM

The colonial hydroid *Hydractinia* is a common inhabitant of nearshore waters in temperate and boreal regions worldwide. In the North Atlantic, *Hydractinia echinata* and related species (Buss & Yund 1989) are

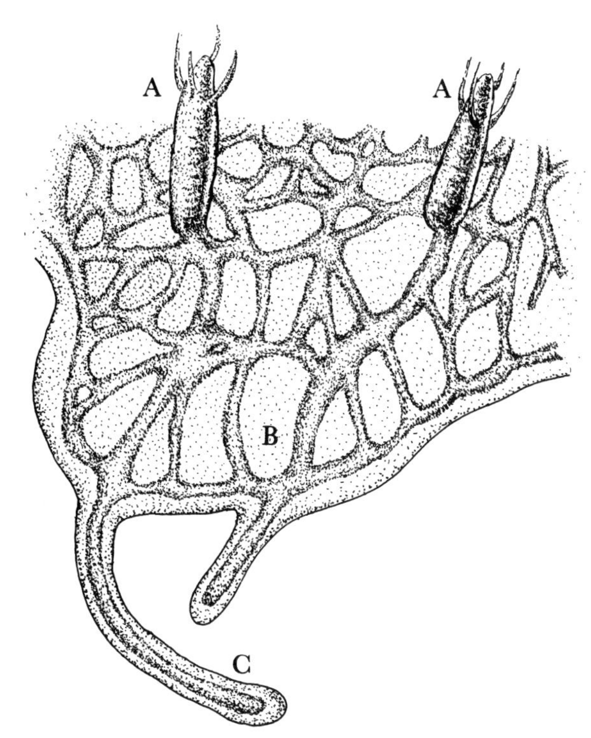

Figure 2. Diagram of a *Hydractinia* colony showing: (*a*) feeding polyps, (*b*) stolonal mat and (*c*) peripheral stolons.

frequently found encrusting gastropod shells inhabited by hermit crabs. The life cycle of *Hydractinia* is simple; dioecious colonies shed gametes, fertilization is external, and the zygote gives rise to a crawling planula larva. The larva settles on a hermit crab shell, metamorphoses into a primary polyp, and grows to give rise to a colony. Asexual propagation from one shell to another is not a common feature of the life cycle.

A *Hydractinia* colony is composed of three distinct structures: polyps, stolonal mat, and peripheral stolons (figure 2). Polyps are feeding organs (gastrozooids), organs of reproduction (gonozooids), and organs of defence (tentacular zooids and dactylozooids). Of these polyp types, only feeding zooids are relevant to the current investigation. Defensive polyps occur only when induced by other organisms, hence do not appear in monospecific laboratory culture. Gonozooids only appear when the limits of a substratum have been reached (or at very large colony size), and these conditions were experimentally precluded. Feeding polyps arise from either stolonal mat or peripheral stolons. Stolonal mat is a closely knit complex of anastomosing endodermal canals capped, except at the mat periphery, with a uniform sheet of ectoderm. Peripheral stolons are continuous with the endodermal tubes of the mat, but extend beyond it.

Growing *Hydractinia* colonies display substantial variation in the relative proportions of these three structures, hence in colony shape (figure 3). Variation ranges from clones that lack peripheral stolons altogether ('mat-type' of Hauenschild 1954) to colonies that bear high proportions of peripheral stolons ('net-type' of Hauenschild 1954). This variation in colony morphology is apparent only in the early stages of ontogeny. As colonies come to cover all available substratum, peripheral stolons develop into stolonal mat. Ontogenetic variation in colony morphology thus largely parallels the between-taxa variation captured by the 'phalanx' versus 'guerilla' distinction (table 1).

Hydractinia is susceptible to the usual range of physical challenges (e.g. desiccation, salinity stress, temperature) and natural enemies (e.g. disease, predators, intra- and interspecific competitors). Of these, only intraspecific competition is known to discriminate between colonies on the basis of colony morphology (Buss *et al.* 1984; Buss & Grosberg 1990). When colonies of *Hydractinia* encounter one another, one of two results may occur; colonies may either fuse or reject. Fusion is limited to close kin and occurs only rarely in natural populations (Hauenschild 1954, 1956; Buss & Shenk 1990). Rejection, too, may take one of two forms – either passive or aggressive (Buss *et al.* 1984; Buss & Grosberg 1990; Buss & Shenk 1990). Passive rejection occurs when two mat-type colonies encounter one another, whereupon both colonies erect a barrier to further cell–cell contact and co-exist. When a net-type colony encounters another colony, an active aggressive response is triggered (Ivker 1972; Buss *et al.* 1984; Lange *et al.* 1989).

The aggressive response has two components: the local proliferation of stolons and the destructive effect of stolons upon one another. Whenever a stolon encounters an incompatible one, a new tip is induced, which may, in turn, induce further tip formation in subsequent encounters (Müller *et al.* 1987). This capacity for local induction of stolonal tips leads to local stolonal proliferation in areas where the stolons of two colonies come into contact. In addition to local proliferation, stolon tips differentiate in the presence of conspecifics. Ultrastructural (Buss *et al.* 1984) and time-lapse video analysis (Lange *et al.* 1989) have documented the migration of nematocytes into stolonal tips, producing swollen (i.e. hyperplastic) stolons. Hyperplastic stolons discharge their nematocysts into neighbouring tissues and effect local destruction (Buss *et al.* 1984; Lange *et al.* 1989). This process continues at points of intraspecific stolonal contact until one colony has annihilated the other, or until all stolons disappear and colonies co-exist.

Since net-type colonies are superior intraspecific competitors, intraspecific variation in *Hydractinia* differs from the usual pattern in this life-history correlate (cf. table 1). Further, since only stolons are capable of giving rise to hyperplastic stolons (Buss *et al.* 1984; Buss & Grosberg 1990), net-type colonies gain this superiority by virtue of their greater amounts of peripheral stolons. Given the multiplicative nature of the tip proliferation, it is not surprising that the absolute amount of peripheral stolon tissue has been found to be an important predictor of competitive ability (Buss & Grosberg 1990).

Several lines of evidence suggest that selection on colony morphology occurs in natural populations:

1. *Variation in morphologically based competitive ability exists and is heritable.* In natural populations, a continuous range of variation, from extreme mat-type to extreme net-type colonies, is found. Analysis of clonal

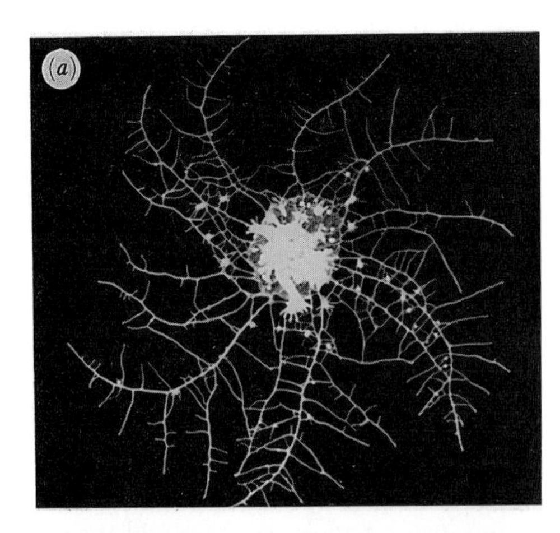

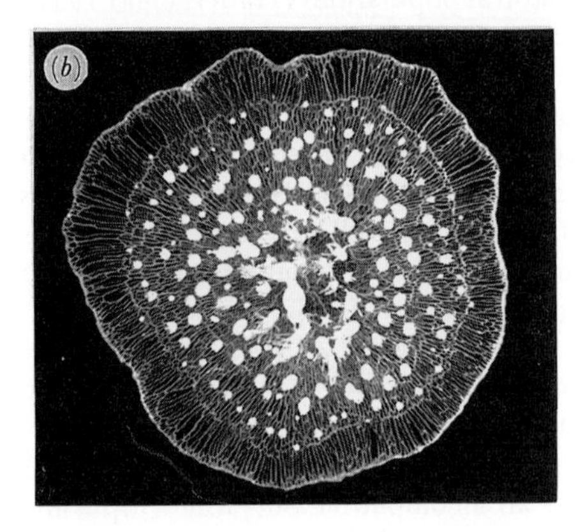

Figure 3. Photographs of a (*a*) 'net-type' and a (*b*) 'mat-type' colony of *Hydractinia*.

repeatibility (i.e. 'common garden experiments') and breeding studies have shown that variation in colony morphology exhibits broad-sense heritability (McFadden *et al.* 1984; Blackstone & Yund 1989; Buss & Grosberg 1990; Blackstone & Buss, 1991; see Falconer, 1981, for general discussion of heritability).

2. *Intraspecific competition occurs chronically in natural populations.* Studies of nine populations in the north-eastern United States have shown that intraspecific competition is a common occurrence in all populations (Yund *et al.* 1987; Buss & Yund 1988; Yund & Parker 1989). In addition, museum specimens spanning the past century display frequencies of intraspecific competition within the range detected in modern populations (Buss & Yund 1988). Finally, limited samples from Miocene populations display signatures of population structure (e.g. size-frequency distributions) known to be driven by intraspecific competition in modern populations (Buss & Yund 1988).

3. *Within and between population variation suggests selection for competitive ability.* The frequencies of colonies of net-type morphology covaries with the frequency of intraspecific competition in between-population comparisons (Yund 1991). Within a given population, the frequency of net-type colonies is enhanced on completely covered shells, which have experienced the selective filter, relative to the frequency of newly recruited colonies (Yund 1991).

4. *Morphological specialization.* The occurrence of a morphology, the hyperplastic stolon, that is specialized for competition is strong *a priori* grounds for suspecting a chronic importance of this process. Notable is the fact that the hyperplastic stolons of *Hydractinia* are known to be induced only upon contact with a competitor and that they are effective only against conspecifics (McFadden 1986).

In summary, *Hydractinia* displays intraspecific variation in colony morphology that mimics the variation displayed between higher taxonomic groups, and this variation is believed to be subject to natural selection in modern and historical populations. Thus, *Hydractinia* should regulate its morphology when perturbed; in particular, the amount of peripheral stolons should be closely regulated.

3. EXPERIMENTAL STUDIES

(*a*) *Methods*

A single clone of *Hydractinia echinata*, derived from a colony collected from Sylt, North Sea, was used in all studies. Clonal explants were made from stock colonies by standard techniques of explanting (McFadden *et al.* 1984). Explants are small regions of stolonal mat (less than 2 mm^2) bearing 3–5 polyps surgically removed from stock colonies. Explants were held affixed to glass microscope slides with a loop of quilting thread. In 2–4 days, explants attached to the slides and the thread was removed. Twenty clonal replicates established in this manner were maintained in 170-gallon (773 l) re-circulating aquaria at 15 °C. Colonies were fed to repletion daily with four-day-old *Artemia* nauplii.

All replicates were observed daily with a Wild dissecting scope and the number of polyps counted directly. The outline of the stolonal mat of each colony was traced from a camera lucida projection. In addition, the extent of peripheral stolon development was estimated by determining the area of the complex polygon generated by connecting the endpoints of all free stolon tips from the camera lucida tracing. The area enclosed by stolons and the area of the stolonal mat was calculated by manually digitizing the camera lucida tracing using a Summagraphic bitpad interfaced to an Apple Macintosh microcomputer.

Three replicates were designated as controls and allowed to grow to a size of > 500 polyps without experimental perturbation. The remaining 17 replicates were perturbed from their normal growth trajectory and thereafter left unperturbed until the colony reached a size of > 500 polyps. The 500 polyp threshold was chosen to substantially exceed the size at which colonies experience high frequencies of intraspecific competition in natural populations (Yund *et al.* 1987; Buss & Yund 1988; Yund & Parker 1989). Perturbations were effected by surgically removing polyps, peripheral stolons or stolonal mat with a microscalpel. Perturbations were chosen in attempt to sample as fully as practical the entire range of mat–polyp–stolon 3-space.

Assessment of the effects of surgery per se is critical to the interpretation of these experiments. *Hydractinia* repairs surgical injury within seconds. The endoderm of the stolonal vasculature is populated by contractile vacuoles (Schierwater *et al.* 1991), which permit rapid contraction of stolons at the site of injury. To insure, however, that the subsequent growth trajectory of a manipulated colony was not governed by the manipulation itself, we perturbed two replicates to point in the mat–polyp–stolon 3-space within the range of variation displayed by unmanipulated control colonies. If surgical effects governed subsequent growth, then these colonies may be expected to deviate from that observed for control colonies.

(*b*) *Analysis*

Operationally, we defined regulation as the response of a manipulated colony which differs from that of a control colony and serves to return the manipulated colony to within the range of the control growth trajectory. The experimental protocol perturbed the growth of the manipulated colonies in state space. In response, the manipulated colony could (i) grow all structures, manipulated and unmanipulated, at the same rates as the unperturbed, control colonies; (ii) grow the manipulated (reduced) structure(s) at an accelerated rate compared to the control colonies, or (iii) grow the reduced structure(s) at an accelerated rate and the unmanipulated structure(s) at a reduced rate compared to the control colonies. Using our operational definition, (ii) or (iii) would constitute regulation.

If regulation occurs, it can produce convergence of control and manipulated growth trajectories in rate or in state space. The latter can be assessed by comparing growth trajectories for structures of control and manipulated colonies on arithmetic axes. Similarly, regulation of rates can be assessed by comparing growth trajectories for control and manipulated colonies on log-transformed axes, in which slopes represent the relation between the specific growth rates of the structures in question (this follows from the derivative form of the allometry equation; for examples, see Blackstone (1987), Blackstone & Yund (1989)).

Specific growth rates were used as measures of growth rate. Each rate was calculated as increment in area or number (area for stolonal mat and peripheral stolon, number for polyp) per time increment per initial area or number. Although 'specific' technically refers to 'divided by mass', any measure of size can be used, as a specific growth rate has units of 1/time. This rate approximates the derivative with respect to time of the logarithm of the size measure and is a physiologically meaningful measure of rate (see, for example, Stebbing 1981). For each of the ten days following the surgical manipulation, specific growth rates for polyps, stolonal mat, and peripheral stolons were calculated (in some cases data were available for 11 daily intervals). Thus, the growth of a single colony was assessed by ten (or in some cases 11) specific growth rates. Using analysis of variance, the rates of stolon, mat, and polyp production were compared between control and manipulated colonies; using multivariate analysis of variance (MANOVA), the relations among these rates were compared between control and manipulated colonies.

Growth trajectories were compared visually and with analysis of covariance; because competition focuses on the area of peripheral stolons, this area was treated as the outcome variable. The correlations of polyp, mat, and stolon generally exceed 0.95; this colinearity rendered multiple regression of little use.

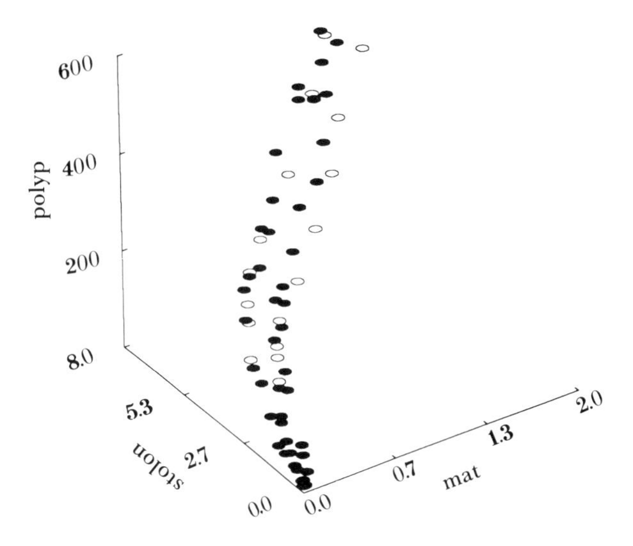

Figure 4. Three-dimensional scatterplot of polyp number, stolonal mat and peripheral stolon area (cm^2) for unmanipulated colonies (closed circles) and colonies manipulated to approximate points in the control growth trajectory (open circles).

(*c*) *Results*

We do not attempt here to utilize the growth trajectories of perturbed colonies to construct equations of state for developing *Hydractinia* colonies. Rather we will seek only to provide a qualitative description of the behaviour of perturbed colonies. We will ask only whether there is evidence for regulation and convergence of rates, states, or both, between control and manipulated colonies.

Controls. For two control manipulations (i.e. colonies that were surgically manipulated, but not displaced from the control trajectory in state space) and two unperturbed colonies over the ten-day period following the manipulations, there was no significant difference in the rate of production of polyps ($F = 2.38$, d.f. = 1,38, $p > 0.10$), stolonal mat ($F = 0.98$, d.f. = 1,38, $p > 0.30$), or peripheral stolons ($F = 0.03$, d.f. = 1,38, $p > 0.80$), and the relation among these rates does not differ between control and manipulated colonies (MANOVA, approximate $F = 1.23$, d.f. = 3,36, $p > 0.30$). The growth trajectories of unmanipulated and control colonies are shown in figure 4.

Regulation. Perturbed *Hydractinia* colonies exhibit regulation of specific growth rates. Examples of this regulation will be provided for three representative classes of manipulations performed after two weeks of growth.

1. Peripheral stolon area reduced (50%), mat and polyps undisturbed. Polyp growth was unchanged ($F = 0.05$, d.f. = 1,19, $p > 0.80$), stolonal mat growth was unchanged ($F = 0.79$, d.f. = 1,19, $p > 0.35$), while peripheral stolon growth was accelerated ($F = 6.94$, d.f. = 1,19, $p < 0.05$) in one manipulated colony

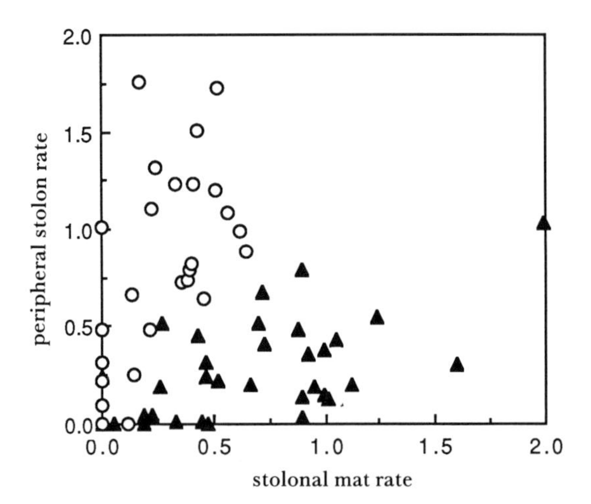

Figure 5. Bivariate scatterplot of specific growth rates ($1/10^2$/h) for peripheral stolon and stolonal mat areas from three control colonies (triangles) and three colonies in which polyp number and peripheral stolon area were reduced by 50 and 75% respectively (circles). Specific growth rates are for the ten days following the manipulations; while polyp growth rate was unchanged, note that the manipulated colonies exhibit not only an acceleration of peripheral stolon growth, but a slowing of stolonal mat growth.

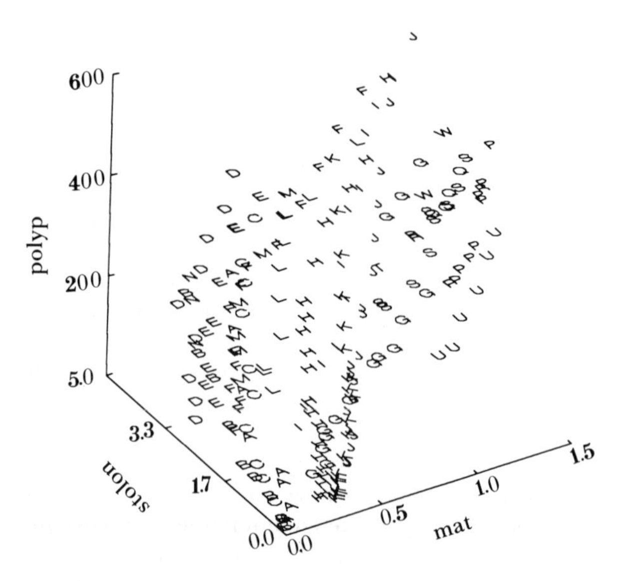

Figure 6. Three-dimensional scatterplot of polyp number, stolonal mat area (in square centimetres) and peripheral stolon area (in square centimetres) for control and all manipulated colonies. Each colony is represented by a different letter and each sequence of the same letter represents the growth trajectory of that colony at daily intervals.

compared to one unmanipulated control over the same ten-day time period. The relation among rates was altered as well (MANOVA, approximate $F = 6.41$, d.f. = 3,17, $p < 0.01$), although this is likely to reflect the acceleration of the peripheral stolon growth in the manipulated colony rather than more complex regulation.

2. Peripheral stolon area reduced (75%), polyp number reduced (50%) and mat left undisturbed. For three manipulated colonies compared to three unmanipulated controls over the same time period, polyp growth was unchanged ($F = 0.33$, d.f. = 1,60, $p >$ 0.55), stolonal mat growth was slowed ($F = 21.77$, d.f. = 1,60, $p \ll 0.001$), and peripheral stolon growth was accelerated ($F = 19.03$, d.f. = 1,60, $p \ll 0.001$) over the same ten-day time period. A highly significant MANOVA (approximate $F = 25.4$, d.f. = 3,58, $p \ll$ 0.001) reflects the dramatic changes in the relation of stolonal mat and peripheral stolon growth rates in these manipulations (figure 5).

Although polyp growth rate was unchanged over the ten-day period, it should be noted that polyp growth rate was extremely rapid for the 24-hour period immediately following the manipulation (0.027 polyps $\text{polyp}^{-1}\,\text{h}^{-1}$ versus 0.011 for control colonies, $F = 7.74$, d.f. = 1,4, $p < 0.05$). After perturbation, the colony produces new polyp buds, generally in the same location of those removed, and after this period of formation of polyp buds, polyp growth returned to the control value ($F = 2.00$, d.f. = 1,4, $p > 0.20$).

3. Polyp number reduced (> 50%), mat area reduced (> 50%), and peripheral stolons left undisturbed. For two manipulated colonies compared to two unmanipulated controls over the same time period, polyp growth accelerated slightly ($F = 4.21$, d.f. = 1,40, $p < 0.05$), stolonal mat growth accelerated slightly ($F = 5.10$, d.f. = 1,40, $p < 0.05$), whereas peripheral stolon growth was unchanged ($F = 0.08$, d.f. = 1,40, $p > 0.75$). The relation among rates was altered as well (approximate $F = 2.98$, d.f. = 3,38, $p <$ 0.05).

The patterns displayed by these three examples appear to be generally true; all three structures tend to exhibit a regulatory response in growth rate when perturbed. Further, both the reduction of growth of unmanipulated structures as well as the acceleration of growth of manipulated structures can occur (figure 5), suggesting a somewhat complex colony-wide regulation of growth.

Convergence onto control trajectories. Perturbed *Hydractinia* colonies exhibit regulation of rates, but regulation is not sufficiently rapid to return colonies to the control growth trajectory in mat–polyp–stolon space. Figure 6 shows the trajectories of all disturbed colonies. It is apparent that regulation does not result in a rapid attainment of control values. Using the representative manipulations introduced above, we examine the relation between control and perturbed colonies in more detail. For the ranges relevant to this analysis, the relations between both arithmetic and log-transformed values of stolonal mat and peripheral stolon areas are linear, at least to a first approximation. Thus, for both rates and states, differences between control and manipulated trajectories can be examined using analysis of covariance. In particular, convergence of trajectories is apparent from heterogeneity of slopes, i.e. the presence of a significant interaction in the analysis.

1. Peripheral stolon area reduced (50%), mat and polyps undisturbed. In this case, the arithmetic values (peripheral stolon area as the outcome, stolonal mat area as the covariate) show a non-significant interaction between control and manipulated colonies ($F = 0.43$, d.f. = 1,22, $p > 0.50$), but nevertheless the control and manipulated trajectories differ strongly in

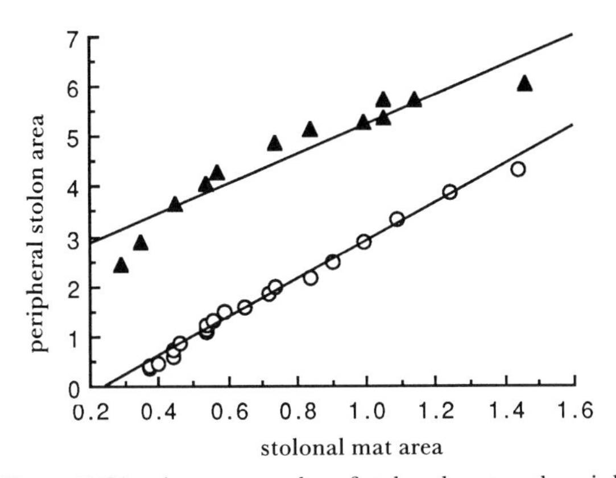

Figure 7. Bivariate scatterplot of stolonal mat and peripheral stolon area (in square centimetres) for a control colony (triangles) and a colony in which stolonal area was reduced (circles). Lines represent least-squared regressions.

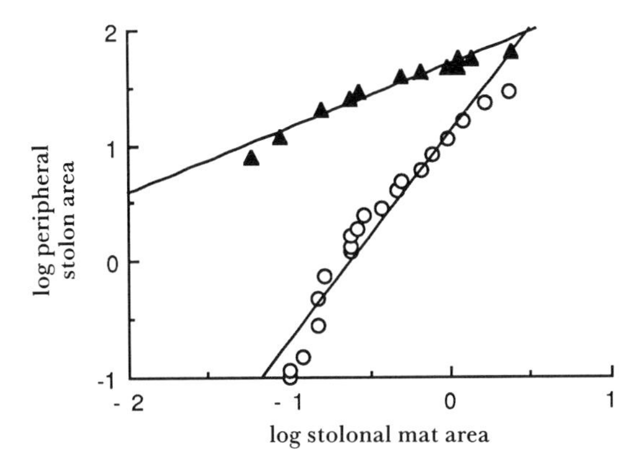

Figure 8. Bivariate scatterplot of log-transformed values of stolonal mat and peripheral stolon area for the control colony (triangles) and the manipulated colony (circles) from figure 7. Lines represent least-squared regressions.

elevation ($F = 286$, d.f. $= 1,23$, $p \ll 0.001$). Thus, these trajectories are distinct and not converging. On the other hand, analysing the log-transformed values shows a highly significant interaction ($F = 57.25$, d.f. $= 1,22$, $p \ll 0.001$), and the control and manipulated trajectories converge by virtue of this difference in slope. Note that this heterogeneity of slopes is analogous to the difference in specific growth rates discussed above (see Blackstone 1987; Blackstone & Yund 1989).

2. Peripheral stolon area reduced (75%), polyp number reduced (50%) and mat left undisturbed. Each of these manipulated colonies are similar to the colony discussed above, e.g. for one colony and one control, figures 7 and 8 show the arithmetic and log-transformed trajectories. For arithmetic values, there is a significant interaction term ($F = 6.24$, d.f. $= 1,29$, $p < 0.05$); for log-transformed values the interaction is highly significant ($F = 101$, d.f. $= 1,29$, $p \ll 0.001$). Although trajectories converge in both rate and state space, convergence of rates is much stronger. Further, this convergence is due to both an increase in the rate of peripheral stolon growth and to a decrease in stolonal mat growth.

3. Polyp number reduced (> 50%), mat area reduced (> 50%) and peripheral stolons left undisturbed. Both of these manipulated colonies exhibit the same patterns as those seen in the above examples; weak regulation of states and strong regulation of rates. For instance, for one of the manipulated colonies and one control, the arithmetic values for peripheral stolon and stolonal mat area show a significant interaction ($F = 5.90$, d.f. $= 1,14$, $p < 0.05$), whereas the log-transformed values show a much stronger interaction ($F = 98$, d.f. $= 1,14$, $p \ll 0.001$).

In summary, the experimental manipulations of *Hydractinia* colonies provide some interesting results; although growth rates of manipulated (reduced) structures are often accelerated (and growth rates of unmanipulated structures are sometimes slowed, figure 5), and although the relations of growth rates rapidly converge on control trajectories (figure 8), this regulation does not produce convergence on control growth trajectories in state space (figures 6 and 7).

4. DISCUSSION

(*a*) *Is the* Hydractinia *result generalizable?*

Our results clearly show that *Hydractinia* exhibits regulation of growth rates, but that regulation is not sufficiently rapid for colonies to re-establish control growth trajectories following disturbance. One might reasonably ask whether this result is likely to be generally true for clonal organisms or whether it is peculiar to this hydroid. Among benthic hydroids, *Hydractinia* has long been recognized as the pinnacle of colony integration (Mackie 1973, 1986). The anatomy of its vascular system is far more sophisticated than that of most other hydroids, in terms of the number and types of stolonal anastomosis that occur and, hence, *Hydractinia* would be among the hydroids most likely to display regulation. Comparisons between taxa are more difficult. Some clonal taxa may well display more effective regulation. Bryozoan zooids communicate through funiculi that are clearly less efficient, in terms of transport rates, than is the colonial vasculature of hydroids (Best & Thorpe 1985). Nevertheless, the rigid zooidal architecture of these forms and the need of some colonies to generate communal feeding currents may limit the range of permissible deviations from a given colony form. In most other clonal groups, however, ramets are only loosely connected by stolons and display far more restricted sharing of resources through ramets than is the case in *Hydractinia* (Pitelka & Ashmun 1985). Control of the distribution of resources is often localized to individual ramets or small numbers of nearby ramets. In any such clone, one may suspect that perturbations of these systems will generate results similar to those we report here.

(*b*) *The application of Waddington's concept to clonal taxa*

Waddington was clearly thinking of aclonal organisms in formulating his conclusions with respect to canalization. This is particularly evident in his concentration on the end-product of a developmental sequence. Unlike aclonal organisms, development in clonal organisms is continuous and few clonal organisms come to adopt a fixed end-product. Perhaps it is a disservice to Waddington's memory to apply his notions to clonal organisms. We think not. The development of any organism can be approached as a dynamical systems problem. More importantly, Waddington's central point is that canalization should occur in any trait subject to selection and this argument is not sensitive to the choice of organism.

(c) *Interpretation of experimental findings*

Given that Waddington's argument is applicable to this case, there are several possible interpretations of our findings.

The adaptation interpretation. In both the modern and historical populations we have studied (Yund *et al.* 1987; Buss & Yund 1988; Yund & Parker 1989; Yund 1991), the vast majority of competitive interactions occur as pairwise encounters. Most encounters occur at small colony size ($\ll 50$ polyps) by virtue of the fact that colonies preferentially recruit to sites on the inner lip of the aperture of a shell. Such encounters are rapidly resolved in favour of one colony or another while colonies are still of small size (Yund *et al.* 1987). Given the dependence of competitive ability on the quantity of stolonal tissue and the multiplicative relation between tip proliferation and initial tip number in aggressive interactions, one would expect that rapid regulation of state would be strongly favoured by selection. Yet, regulation of state was not observed; the observed regulation of rates is insufficient to return a colony to pre-disturbance morphology (figure 6).

Nevertheless, selection may have produced canalized rates of increase of stolons, i.e. hyperplastic stolon production depends on both the initial number and on the rate of tip production. Given the slow convergence of manipulated colonies onto control growth trajectories, a high rate of stolon production will likely be favoured only in the small percentage of competitive interactions that occur at large colony size. Such interactions are found when one colony recruits to the site near the aperture of the shell and another recruits to as site distant from the lip of the aperture, most commonly in the siphonal notch (Yund *et al.* 1987; Buss & Yund 1988; Yund & Parker 1989).

The non-adaptation interpretation. The observed regulation of rates is not sufficiently rapid to support an adaptationist interpretation for the most common selective environment, that of pairwise encounters at small colony size. In such contests, initial stolon number is crucial, since the encounter is too brief for rate of production to compensate for an initial disadvantage. Perhaps *Hydractinia* colonies do not rapidly regulate mat–polyp–stolon state because they cannot; they lack appropriate genetic variation for selection to work upon. In this view, selection is potentially efficacious in all matters except rapid canalization, and whereas the most common selective environment favors rapid regulation of state, this is impossible and the organism persists despite this defect.

The structuralist interpretation. A developing colony may, in principle, be characterized by a series of equations defining state and transitions in state. At present, such descriptions are lacking for any organism. It is reasonable to ask, however, what would be the range of dynamical behaviour displayed by such a description? Goodwin and others (Goodwin 1982; Goodwin & Saunders 1989; Goodwin *et al.* 1989) have noted that it is conceivable that the underlying dynamics of a system may be competent to produce observed patterns without appeal to the action of natural selection. Perhaps it is the case that correlations, say between life-history traits, are *necessarily* produced by features of the underlying dynamics. If this is the case, then the fact that natural selection favors these correlations does not necessarily imply that these correlations have been shaped by natural selection.

Consider the effect of a disturbance on a *Hydractinia* colony. Stolons are essentially tubes for the transport of the hydroplasm. This fluid is propelled by contraction of polyps and stolons elongate most rapidly when flow is directed into their tips (Wyttenbach 1973). Polyps are connected at their bases to several stolons. A particular polyp may be connected to both narrow stolons of the stolonal mat and wide stolons leading through the stolonal mat to peripheral stolons. In one of our representative examples, disturbances were made to both peripheral stolons and polyps. The removal of polyps resulted in a replacement by new, smaller polyps. Thus, the capacity of polyps to propel a given volume of fluid is reduced after a disturbance. Because peripheral stolons are wider than stolons of the mat, these stolons will receive a greater volumetric rate of flow than stolons in the mat, because flow rate is proportional to the radius. Thus, after the disturbance the peripheral stolons will elongate at faster absolute rates than do stolons in the mat. As a purely physical consequence, then, one would expect that the relative rate of stolon growth will be accelerated and rate of mat growth be depressed in a disturbance of this sort. This pattern of 'regulation' is just what is observed in such manipulations (figure 5).

Although we are far from a dynamical system portrait of the hydrodynamics of colony form in *Hydractinia*, such a description may generate the pattern of regulation of rates that we have observed. Should this be the case, one interpretation of our experimental findings is that these hydrodynamical considerations are but the proximate basis for the ultimate pattern produced by selection. The structuralist interpretation is not, however, a claim about proximate explanations; it is a claim that ultimate explanations are unnecessary for certain classes of proximate claims. The general distinction may be stated in terms of the *Hydractinia* problem. Imagine that the equations defining hydroid growth in terms of the underlying hydrodynamics provide multiple solutions; say, colonies may display either regulation of rates or no regulation of rates. Then the claim that selection has chosen among these solutions to favour regulation of rates is necessary. If, however, one finds that the *only* solution to equations defining these hydrodynamics is that of regulation of rates, then a claim for the efficacy of selection is neither necessary nor sufficient.

The problem here is one of null models. Selection is a process that acts upon existing variation. We do not know the generating functions that produce that variation and can only gain a highly restricted glimpse of the dynamical potential by assessment of existing variation. Hence to rigorously assess whether selection is efficacious, or even necessary, we must first know the range of permissible dynamics. These must constitute the null model against which we assess the role of selection. Note that this structuralist position is not at variance with the non-adaptation interpretation, but that its claims are more pervasive. The non-adaptation interpretation permits selection to be efficacious in all cases in which appropriate genetic variation occurs. The structuralist perspective is more demanding in that it requires first a complete understanding of the range of possible dynamical behaviours before assessment can be made of whether selection is necessary to account for patterns we observe.

With our present understanding of this system, we cannot say which of these three interpretations is correct, or even most likely to be correct. We do, however, strongly advocate the view that the latter two interpretations are as worthy of attention as the former.

This work was motivated by discussions with J. Rimas Vaisnys. We thank J. Bartels, E. Beeler, D. Bridge, C. Cunningham, R. DeSalle, M. Dick, J. Reinitz, B. Schierwater and L. Vrba for comments, J. Taschner for technical assistance, and NSF (BSR-8805961) and ONR (N00014-89-J-3046) for support.

REFERENCES

Bell, A. D. 1984 Dynamic morphology: a contribution to plant population ecology. In *Perspectives on plant population ecology* (ed. R. Ditzo & J. Sarukán), pp. 48–65. Sunderland, Massachusetts: Sinauer Associates.

Best, M. A. & Thorpe, J. P. 1985 Autoradiographic study of feeding and the colonial transport of metabolites in the marine bryozoan *Membranipora membranacea*. *Mar. Biol.* **84**, 295–300.

Blackstone, N. W. 1987 Specific growth rates of parts in a hermit crab: a reductionist approach to the study of allometry. *J. Zool., Lond. A* **211**, 531–545.

Blackstone, N. W. & Yund, P. O. 1989 Morphological variation in a colonial marine hydroid: a comparison of size-based and age-based heterochrony. *Paleobiology* **15**, 1–10.

Blackstone, N. W. & Buss, L. W. 1991 Shape variation in hydractiniid hydroids. *Biol. Bull.* **180**. (In the press.)

Boardman, R. S., Cheetam, A. H. & Oliver, W. A. 1973 *Animal colonies*. Stroudsburg, Pennsylvania: Dowden, Hutchinson and Ross.

Buss, L. W. 1979 Habitat selection, directional growth, and spatial refuges: why colonial animals have more hiding places. In *Biology and systematics of colonial organisms* (ed. G. Larwood & B. Rosen), pp. 459–477. London: Academic Press.

Buss, L. W. 1987 *The evolution of individuality*. Princeton University Press.

Buss, L. W. & Grosberg, R. K. 1990 Morphogenetic basis for phenotypic differences in hydroid competitive behaviour. *Nature, Lond.* **343**, 63–66.

Buss, L. W., McFadden, C. S. & Keene, D. R. 1984 Biology of hydractiniid hydroids. 2. Histocompatibility effector system/competitive mechanism mediated by nematocyst discharge. *Biol. Bull.* **167**, 131–158.

Buss, L. W. & Shenk, M. A. 1990 Hydroid allorecognition regulates competition at both the level of the colony and the level of the cell lineage. In *Defense molecules* (ed. J. J. Marchalonis & C. Reinisch), pp. 85–106. New York: Wiley–Liss.

Buss, L. W. & Yund, P. O. 1988 A comparison of modern and historical populations of the colonial hydroid *Hydractinia*. *Ecology* **69**, 646–654.

Buss, L. W. & Yund, P. O. 1989 A sibling species group of *Hydractinia* in the northeastern United States. *J. mar. biol. ass. U.K.* **69**, 857–875.

Caswell, H. 1989 *Matrix population models*. Sunderland, Massachusetts: Sinauer Associates.

Falconer, D. S. 1981 *Introduction to quantitative genetics*. London: Longman.

Goodwin, B. C. 1982 Development and evolution. *J. theor. Biol.* **97**, 43–55.

Goodwin, B. C. & Saunders, P. 1989 *Theoretical Biology: epigenetic and evolutionary order from complex systems*. Edinburgh University Press.

Goodwin, B. C., Sibatani, A. & Saunders, P. T. 1989 *Dynamic structures in biology*. Edinburgh University Press.

Harper, J. L. 1977 *The population biology of plants*. New York: Academic Press.

Harper, J. L. 1985 Modules, branches, and the capture of resources. In *Population biology and evolution of clonal organisms* (ed. J. B. C. Jackson, L. W. Buss & R. E. Cook), pp. 1–33. New Haven, Connecticut: Yale University Press.

Harper, J. L. 1986 *Modular organisms: case studies of growth and form*. London: Royal Society.

Harper, J. L., Rosen, B. R. & White, J. 1986 *The growth and form of modular organisms*. London: Royal Society.

Hauenschild, V. C. 1954 Genetische and entwicklungphysiologische Untersuchungen uber Intersexualitat und bewebevertraeglichkeit bei *Hydractinea echinata* Flem. *Arch. Entwicklungsmech. Org.* **147**, 1–41.

Hauenschild, V. C. 1956 Uber die Vererbung einer Gewebevertraglichkeitseigenschaft bei dem Hydroidpolypen *Hydractinea echinata*. *Z. Naturforsch.* **11b**, 132–138.

Ivker, F. B. 1972 A hierarchy of histoincompatibility in *Hydractinia echinata*. *Biol. Bull.* **143**, 162–174.

Jackson, J. B. C. 1979 Morphological strategies of sessile animals. In *Biology and systematics of colonial organisms* (ed. G. Larwood and B. Rosen), pp. 499–556. London: Academic Press.

Jackson, J. B. C., Buss, L. W. & Cook, R. E. 1985 *Population biology and evolution of clonal organisms*. New Haven, Connecticut: Yale University Press.

Lange, R., Plickert, G. & Müller, W. A. 1989 Histocompatibility in a low invertebrate. *Hydractinia echinata*: analysis of the mechanism of rejection. *J. exp. Zool.* **249**, 284–292.

Larwood, G. & Rosen, B. R. 1979 *Biology and systematics of colonial organisms*. London: Academic Press.

Lovett-Doust, L. 1981 Population dynamics and local specialization in a clonal perennial (*Ranunculus repens*). I. The dynamics of ramets in contrasting habitats. *J. Ecol.* **69**, 743–755.

Mackie, G. O. 1973 Coordinated behavior in hydrozoan colonies. In *Animal colonies* (ed. R. S. Boardman, A. H. Cheetam & W. A. Oliver, Jr), pp. 95–106. Stroudsburg, Pennsylvania: Dowden, Hutchinson and Ross.

Mackie, G. O. 1986 From aggregates to integrates: physiological aspects of modularity in colonial animals. *Phil. Trans. R. Soc. Lond.* B **313**, 175–196.

McFadden, C. S. 1986 Laboratory evidence for a size-refuge in competitive interactions between the hydroids

Hydractinia echinata (Fleming) and *Podocoryne carnea* (Sars). *Biol. Bull.* **171**, 161–174.
McFadden, C. S., McFarland, M. & Buss, L. W. 1984 Biology of hydractiniid hydroids. 1. Colony ontogeny in *Hydractinia echinata. Biol. Bull.* **166**, 54–67.
McKinney, F. & Jackson, J. B. C. 1989 *Bryozoan evolution.* Boston: Unwin Hyman.
Müller, W. A., Hauch, A. & Plickert, G. 1987 Morphogenetic factors in hydroids: 1. Stolon tip activation and inhibition. *J. exp. Zool.* **243**, 111–124.
Pitelka, L. F. & Ashmun, J. W. 1985 Physiology and integration of ramets in clonal plants. In *Population biology and evolution of clonal organisms* (ed. J. B. C. Jackson, L. W. Buss & R. E. Cook), pp. 399–436. New Haven, Connecticut: Yale University Press.
Salzmann, A. G. 1985 Habitat selection in a clonal plant. *Science, Wash.* **228**, 603–604.
Schierwater, B., Piekos, B. & Buss, L. W. 1991 Hydroid stolonal contractions mediated by contractile vacuoles. *J. exp. Biol.* (In the press.)
Schmid, B. 1985 Clonal growth in grassland perennials. III. Genetic variation and plasticity within populations of *Bellis perennis* and *Prunella vulgaris. J. Ecol.* **73**, 819–830.
Schmid, B. 1986 Spatial dynamics and integration within clones of grassland perennials with different growth form. *Proc. R. Soc. Lond.* B **228**, 173–186.
Slade, A. J. & Hutchings, M. J. 1987 Clonal growth and plasticity in foraging behavior in *Glechoma hederacea. J. Ecol.* **75**, 1023–1036.
Stearns, S. C. 1976 Life-history tactics: a review of the ideas. *Q. Rev. Biol.* **51**, 3–47.
Stearns, S. C. 1977 The evolution of life-history traits: a critique of the theory and a review of the data. *A. Rev. Ecol. Syst.* **8**, 145–171.
Stebbing, A. R. D. 1981 The kinetics of growth control in a colonial hydroid. *J. mar. biol. Ass. U.K.* **61**, 35–63.
Sutherland, W. J. 1987 Growth and foraging behavior. *Nature, Lond.* **330**, 18–19.
Sutherland, W. J. & Stillman, R. A. 1988 The foraging tactics of plants. *Oikos* **52**, 239–244.
Waddington, C. H. 1942 Canalization of development and the inheritance of acquired characteristics. *Nature, Lond.* **150**, 563–565.
Wyttenbach, C. R. 1973 The role of hydroplasmic pressure in stolonic growth movements in the hydroid, *Bougainvilla. J. exp. Zool.* **186**, 79–90.
Yund, P. O. 1991 Natural selection on colony morphology by intraspecific competition. *Evolution.* (In the press.)
Yund, P. O., Cunningham, C. W. & Buss, L. W. 1987 Recruitment and post-recruitment interactions in a colonial hydroid. *Ecology* **68**, 971–982.
Yund, P. O. & Parker, H. 1989 Population structure of the colonial hydroid *Hydractinia* sp. nov. C in the Gulf of Maine. *J exp. mar. Biol. Ecol.* **125**, 63–82.

Evolution of alternative reproductive strategies: frequency-dependent sexual selection in male bluegill sunfish

MART R. GROSS

Department of Zoology, University of Toronto, Toronto, Ontario, Canada M5S 1A1

SUMMARY

This study provides empirical evidence in a wild population for frequency-dependent sexual selection between alternative male reproductive strategies. The bluegill sunfish (*Lepomis macrochirus*) has two male reproductive strategies, cuckolder or parental, used by different males to compete in fertilizing the same eggs. As the density of cuckolders in colonies of parental males increases, the average mating success of cuckolders initially peaks but then declines. The cuckolder density at which their success peaks is determined by ecological characteristics of each colony. A theoretical analysis assuming random and omniscient cuckolder distributions among ecologically different colonies shows that cuckolders will fertilize decreasing proportions of eggs, relative to parental males, as cuckolders increase in frequency in the population. This supports evolutionary models that assume negative frequency-dependent selection between the competing strategies. Cuckolder and parental strategies may therefore have evolved as an Evolutionarily Stable State (ESSt).

1. INTRODUCTION

Polymorphisms in male behaviour and life history are found in many species where males compete for access to mates. Examples include antlered and antlerless red deer (Darling 1937), lekking and satellite ruff (van Rhijn 1983; Lank & Smith 1987), territorial and streaker wrasse (Warner 1984), parental and cuckolder sunfish (Gross 1982), and adult and precocious salmon (Jones 1959; Gross 1985; Maekawa & Onozato 1986). In many of these polymorphisms, one phenotype is specialized in fighting for mates while the alternative phenotype, evolved secondarily, is specialized in sneaking. This paper addresses the evolutionary stability of such reproductive polymorphisms.

For a reproductive polymorphism to be evolutionarily stable, a mechanism must exist for the alternatives to have equal fitnesses. Theoretical research suggests that alternative reproductive strategies may evolve through negative frequency-dependent sexual selection (Gadgil 1972; Gross & Charnov 1980; Charnov 1982; Maynard Smith 1982; Gross 1984, 1991; Parker 1984). Imagine that a strategy's mating success depends on its frequency in the population. When rare, the alternative strategy has greater success than the primary strategy, and therefore invades. However, if the alternative strategy's mating success is negatively frequency-dependent (Partridge & Hill 1984; Knoppien 1985), such that with increasing frequency its fitness declines *relative* to that of the primary strategy, an evolutionarily stable frequency may exist where both strategies have equal fitnesses.

At present there is no empirical demonstration of frequency-dependent sexual selection for a wild population (see Partridge 1988). Instead, research has focused on documenting density-dependent selection, usually in an artificial environment. For example, many salmonid species have males that mature at different ages, with older males fighting for mates while younger males sneak matings (Gross 1984). Using electrophoresis, Maekawa & Onozato (1986) and Hutchings & Myers (1988) provide empirical evidence that the success of sneaking males decreases with their density at the oviposition site. Thus with increasing numbers of sneakers, the absolute success per sneaker decreases. Such negative density dependence is not equivalent, however, to negative frequency-dependent selection. The latter requires a decline in the average success per sneaker phenotype *relative* to the average success per fighter phenotype. For example, if a sneaker fertilizes 10% of the eggs when alone with a fighter male at the oviposition site, but only 6% when in a group of ten sneakers, success is negatively density dependent. The fighter male in this example will fertilize 90% and 40% (100% − (10 × 6%)) of the eggs respectively. Therefore the average success per sneaker phenotype relative to that per fighter phenotype has actually increased from 11% (10%/90%) to 15% (6%/40%). This example illustrates that the evolutionary fitness of sneakers may increase with their frequency even though their absolute success declines with their density.

It is important to consider mating sites as subunits from which the cumulative success of strategies is determined at the population level. Subunits will often differ from each other ecologically, and thus the relation between mating success and density may not be constant. Subunits have not been considered in previous studies because manipulations were carried out in artificial environments. The purpose of this paper, therefore, is to outline a test for negative frequency-dependent sexual selection in a wild population. It also presents evidence for both density-dependent and frequency-dependent mating success while incorporating ecological variation in breeding sites. The alternative reproductive strategies studied here are those of the cuckolder and parental males in bluegill sunfish, *Lepomis macrochirus*.

Alternative strategies in male bluegill sunfish

The reproduction of bluegill, an endemic freshwater species of North America, has been studied for 17 consecutive years in Lake Opinicon, near Kingston, Ontario, Canada (see Gross 1979, 1980, 1982, 1984, 1991; Gross & Charnov 1980; Kindler *et al.* 1989). These studies provide ample data for outlining the breeding dynamics of a male polymorphism involving 'parentals' and 'cuckolders.'

Parental male bluegill delay maturity until age seven or eight years. During the summer breeding season, parental males fight among themselves for space to construct a nest within a developing colony. These nests are shallow depressions in the lake bottom made by sweeping motions of the caudal fin. The resulting colonies vary in water depth and cover provided by vegetation and debris. Females arrive at the colony in a school, and the parental males court and subsequently spawn with them. Spawning involves a female entering a nest and repeatedly releasing a small batch of eggs (about 12) using a characteristic dipping motion. The male paired with her fertilizes the eggs during each dip. A male may receive eggs from many different females, accumulating some 30000 eggs in his nest. A female may also spawn in many nests. Spawning within the colony occurs quickly, usually finished within a day. Females then leave the colony while the parental males remain at their nests to provide the care necessary for brood survival. Owing to the dynamics of female arrival, males within a colony and colonies within the lake are highly synchronized in spawning behaviour.

Cuckolder males have a different life history. They mature precociously, usually at age two. Rather than build nests, they distribute themselves among colonies during parental nest-building. When the females arrive, the cuckolders move among the nests within a colony tracking spawning opportunities. The smallest cuckolders behave as sneakers, penetrating the parental male's defence of his nest boundary, and spawning directly over the eggs during the female's dip. Larger cuckolders, four or five years old and about the size of females, act as satellites. These satellites hover in the colony, and follow true females into the nest by mimicking female behaviour. Success depends on deceiving the parental male, and pairing with the female during her dips. Both sneaking and mimicry are ontogenetic tactics within a distinct cuckolder strategy; these males do not live beyond age six. Although only 11–31% of bluegill males in the population mature as cuckolders, the earlier maturity and therefore higher probability of surviving to breed results in mature cuckolders outnumbering mature parental males by a ratio of approximately six to one.

Unlike salmon where several males simultaneously release large clouds of sperm over a single batch of eggs (Jones 1959), sperm competition is probably not a major factor in determining mating success between the parental and cuckolder strategies. Instead, mating success is determined by the behavioural ability to pair with the female during her dip – called 'pairing success'. For parental males, pairing success is determined by their ability to attract females to their nest and then control access to the dips. This control is achieved through guarding against cuckolders by patrolling the nest boundary to detect and chase sneakers and by screening female-sized individuals to detect and rout satellites. Parental males can also control the female dipping behaviour by biting the female. In so doing, the parental risks the female leaving the nest to spawn elsewhere. For cuckolders, pairing success depends upon circumventing the parental male's defence and pairing with the female, quickly placing sperm over the eggs. Both sneakers and satellites use surrounding vegetation and debris as cover from detection and chases by parental males. Cover is also used to avoid piscivorous predators that enter the colonies to feed on the smaller-bodied cuckolders.

In a study of pairing success at seven colonies (Gross 1982), cuckolders attempted to intrude into the nests during nearly 60% of the female dips. Most of these attempts were blocked by parental males, and only 14% of all female dips were successfully paired by cuckolder males. Females did not avoid cuckolder males, but rather spawned readily with them even while the parental male was chasing egg predators or other cuckolders. Thus the primary determinant of cuckolder pairing success was avoiding the parental male. This ability varied among colonies, and was affected by ecological factors as well as interactions among cuckolders themselves. As a consequence, the pairing success at the seven colonies ranged from about 3% to 34% of female dips.

Gross & Charnov (1980) and Gross (1982) proposed that the cuckolder and parental strategies in bluegill may coexist as an Evolutionarily Stable State (ESSt), where both strategies have equal lifetime fitnesses at the balance point. An important assumption in their ESSt models is that strategy fitnesses are regulated by negatively frequency-dependent sexual selection. This would occur if: (i) the pairing success of cuckolders decreases with their density within colonies; and (ii) the distribution of cuckolders among colonies, combined with the density dependence within colonies, results in the relative pairing success of cuckolders to parentals decreasing with increasing cuckolder frequency in the population.

In this paper, cuckolder density is experimentally manipulated at nests within four natural colonies and

Table 1. *The four colonies in Pen Bay*

(Data are mean ± s.e. unless otherwise indicated. The percent distribution of parental males among the four colonies is given.)

variable	colony A	B	C	D
water depth cm	51.2 ± 2.6	27.8 ± 2.6	60.0 ± 5.9	97.2 ± 4.0
cover %	87.9 ± 2.6	55.0 ± 5.0	18.7 ± 4.3	25.8 ± 2.9
parental males n (%)	12 (19.4)	14 (22.6)	12 (19.4)	24 (38.7)
experimental range in cuckolder density per nest	1–11	1–10	1–9	1–12
dips observed:				
total	408	304	243	723
per cuckolder density	40.8 ± 6.9	33.8 ± 9.1	30.4 ± 6.7	60.3 ± 6.8

their pairing success is quantified to examine negative density dependence within the colonies. Since cuckolder distribution among colonies is unknown, two theoretical distributions – random and omniscient – are used to test the density dependence results for negative frequency-dependent mating success at the population level. The random distribution is a conservative assumption that cuckolders do not adjust their numbers among colonies to maximize mating success. The omniscient distribution is a liberal assumption that cuckolders can adjust their distribution among colonies to perfectly maximize their possible mating opportunities.

2. STUDY SITE AND METHODS

The study included all four bluegill colonies, A to D, formed within Pen Bay of Lake Opinicon during July 1981. Our studies both before and after this date suggest that the breeding success observed within these colonies is representative of the population at large. The Pen Bay colonies were therefore studied as ecological subunits within the population.

Each colony was ecologically characterized by the amount of cover afforded cuckolders and by water depth. To quantify cover, a 630 cm^2 plexiglass sheet ruled with a black-and-white checkerboard pattern of 1 cm squares was held 15 cm from a nest's edge. A SCUBA diver, looking out from the centre of the nest, counted the number of white squares not blocked by cover, and repeated this count in three other directions from the same nest centre. The proportion of squares obscured by vegetation or debris was converted to percent cover. Six nests chosen randomly within each colony were sampled and their results pooled to estimate the average cover in the colony. Water depth was measured from the surface to each nest edge.

Skin divers at each colony observed the spawning of cuckolders and parentals to quantify pairing success. As in Gross (1982), success was credited to the phenotype, either cuckolder or parental, paired with the female when she dipped. In only 18% of dips were both male types equally paired to the female; success was then randomly assigned.

To test for density-dependent success within each colony, cuckolder densities were experimentally manipulated by selectively removing individuals from nest sites with a dipnet. This resulted in a broad range of cuckolders per nest site within each colony. The number of cuckolders per nest site ranged from 1 to 12, depending on the colony (table 1; no observations were obtained in colonies A, B and C for 4, 7 and 2 cuckolders per nest, respectively). Neither observation nor removal caused any apparent disturbance to spawning activity.

To test for frequency-dependent success in the overall population, the following mathematical treatment was applied to the data on density-dependent pairing success from the four colonies. The average success of the cuckolder strategy within Pen Bay is a function of the sum of cuckolder successes within each colony. The success within each colony depends on cuckolder density and is therefore a function of the total number of cuckolders in the population and their distribution among the colonies. Thus, the average success of the cuckolder strategy is calculated from

$$\bar{S}c = \sum_{i=A}^{D} (Pc_i)(\bar{S}c_i), \qquad (1)$$

were Pc_i is the proportion of cuckolders in the population that go to the ith colony, and $\bar{S}c_i$ is the average success per cuckolder in the colony. Similarly, the average success of the parental strategy is

$$\bar{S}p = \sum_{i=A}^{D} (Pp_i)(100\% - \bar{S}c_i), \qquad (2)$$

where Pp_i is the proportion of parentals in the population that go to the ith colony. Therefore, the average pairing success of the cuckolder strategy ($\bar{W}c$) *relative* to the average pairing success of the parental strategy ($\bar{W}p$) for different frequencies of cuckolder and parental males in the population is

$$\bar{W}c/\bar{W}p = (Nc/Np)(\bar{S}c/\bar{S}p), \qquad (3)$$

where N is the number of each kind of male in the population.

Unlike the stationary parental males, whose proportional distribution among colonies is easily quantified (Pp_i), the mobile cuckolders are difficult to follow and only their density at specific nest sites, not their proportional distribution among colonies (Pc_i), can be directly quantified. Therefore, to calculate how cuckolder and parental male success ($\bar{W}c/\bar{W}p$) changes

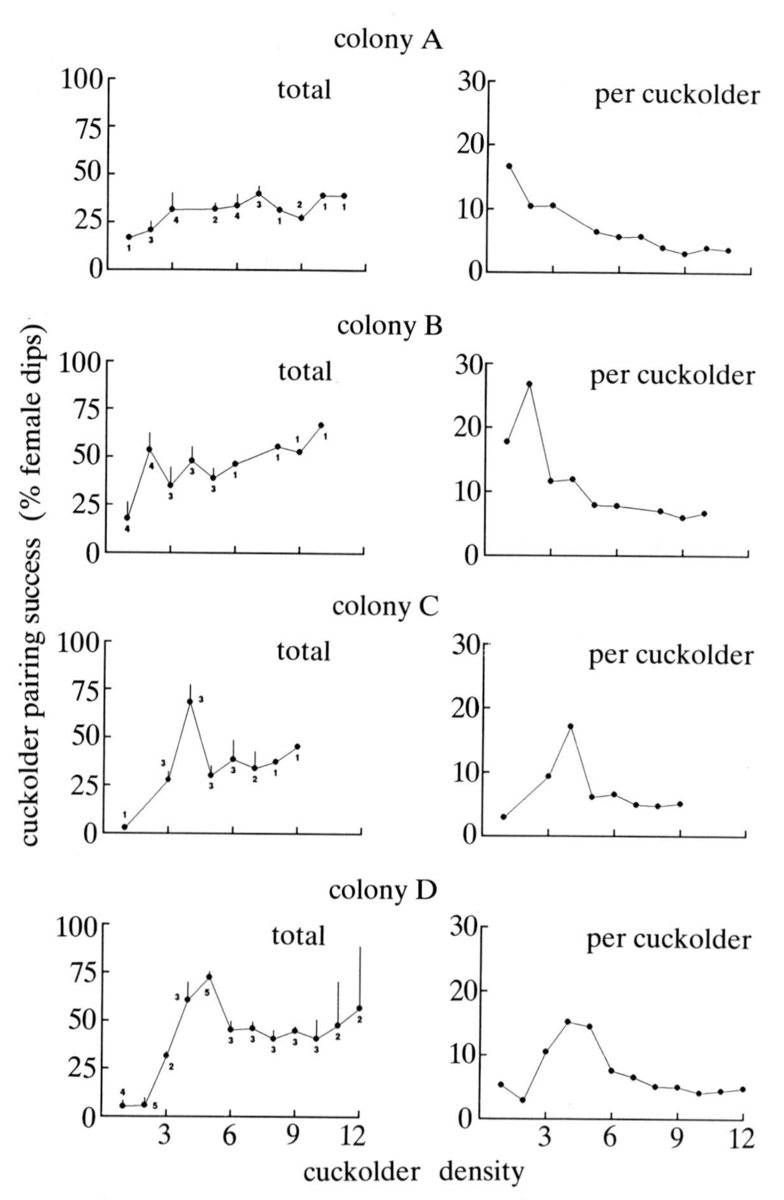

Figure 1. The relation between cuckolder pairing success and cuckolder density (number of cuckolders) at nests in four bluegill colonies, A–D. Both the total success of cuckolders at each density and the average success per cuckolder (total success/number of cuckolders) are plotted. Number of nests sampled at each cuckolder density, and one-half s.e., are shown with the data; individual nests were used more than once when spawning occurred under different cuckolder densities. Percent cover at the colonies is: A = 88 %, B = 55 %, C = 19 %, D = 26 %.

with their relative frequency (Nc/Np), two theoretical distributions – random and omniscient – were used. These theoretical distributions should bracket the true distribution of cuckolders (see, for example, Gross 1984).

In the random distribution, cuckolders disperse among colonies in direct proportion to parental males. Thus, Pc_i is always equal to Pp_i. This assumes that cuckolders do not adjust their distribution among colonies to reflect differences in success with cuckolder density, a conservative assumption that should give a minimum estimate of $\bar{W}c/\bar{W}p$.

In the omniscient distribution, cuckolders possess complete knowledge of how their pairing success varies with density in colonies. They distribute themselves among the colonies such that $\bar{W}c/\bar{W}p$ is always maximized. This assumes that cuckolders approximate an ideal-free mating distribution (Fretwell & Lucas 1970), a liberal assumption as there is both despotic aggression among cuckolders and varying predator pressure among colonies (Gross 1982). The test for negative frequency-dependent selection therefore includes both a minimum (random distribution) and a maximum (omniscient distribution) estimate of relative success ($\bar{W}c/\bar{W}p$) as cuckolder and parental frequency (Nc/Np) change in the population.

3. RESULTS

(a) *Colonies*

The four colonies, located in different areas of Pen Bay, differed ecologically in average water depth and cover afforded to cuckolders (table 1; ANOVA depth, d.f. = 3,20; $F = 155$; $p < 0.001$. ANOVA cover, (% data arcsine square-root transformed) d.f. = 3,20; $F = 68$; $p < 0.001$). However, these variables were not related to the highest natural density of cuckolders observed at a nest in each colony (Spearman rank correlations: depth $r = 0.5$, $p > 0.5$; cover $r = 0.5$, $p > 0.5$). The number of parental males nesting in the colonies was also unrelated to these variables (depth $r = 0.80$, $p > 0.2$; cover $r = 0.4$, $p > 0.5$). Finally, water depth and cover were not significantly correlated ($r = 0.5$, $p > 0.5$).

(*b*) ***Density-dependent pairing success***

Cuckolder pairing success was density dependent within each colony (figure 1). However, the density providing peak average success per cuckolder was negatively correlated with the amount of cover at the colony (Spearman rank correlation, $r = -0.99$, $p < 0.02$). Thus in colony A, with a high of 88% cover, average pairing success per cuckolder peaked at 17% of female dips when only a single cuckolder was present at the nest site and declined to 4% of dips with 11 cuckolders present. By contrast in colony B, with 57% cover, there was an initial increase in average pairing success with density. Here, a single cuckolder achieved 18% of the female's dips, but two cuckolders at the nest each achieved 27%. Success then declined to near 7% with 10 cuckolders present. A similar pattern of initial increase in average pairing success followed by negative density dependence occurred in colonies C and D. However, in these colonies the sparser cover moved the peak yet further to the right.

The decline in average cuckolder success at densities above the peak was related to an increase in the occurrence of simultaneous intrusions, and thus competition among the cuckolder males (cuckolder density versus percent simultaneous intrusions; $n = 9$, $r = 0.894$, $p < 0.01$). For example, at the density of peak success in each of the four colonies, cuckolders competed among themselves for the same dip in only 7% of pairings on average. Beyond this peak, simultaneous intrusions in the four colonies increased to 28% of pairings. At the highest cuckolder densities seen in the colonies, as many as eight individuals were observed entering the nest to fertilize the same dip. Thus the cuckolder males at high densities experienced stronger competition among themselves, than with the parental male. Aggression also increased significantly among cuckolders when their density exceeded peak success within the colonies (chases per cuckolder per minute relative to number of cuckolders present; before peak $= 0.3 \pm 0.3$ s.e., after peak $= 1.2 \pm 0.2$, $t = 4.8$, $p < 0.01$; data pooled over colonies B, C and D, $n = 97$).

Parental males directed aggression not only towards cuckolders, chasing them from the nest area, but also towards the spawning female. Bites to the female slowed her dipping rate (female dips per minute without male aggression $= 5.8 \pm 0.6$ s.e., $n = 51$; with aggression $= 4.8 \pm 0.4$, $n = 28$; $t = 4.9$, $p < 0.05$), and the relative pairing success of parental males increased (pairing success at 5.8 dips per minute = 45%; at 4.8 dips = 82%; $n = 79$, $\chi^2 = 10.26$, $p < 0.01$). However, aggression also increased the probability of females abandoning the nest (bites to female per minute versus total minutes female spawned; $n = 23$, $r = -0.728$, $p < 0.001$). At the highest cuckolder densities, parental males were observed to escalate their aggression to a level where the female was driven from the nest. Cuckolders at such a nest then dispersed to nests with fewer cuckolders.

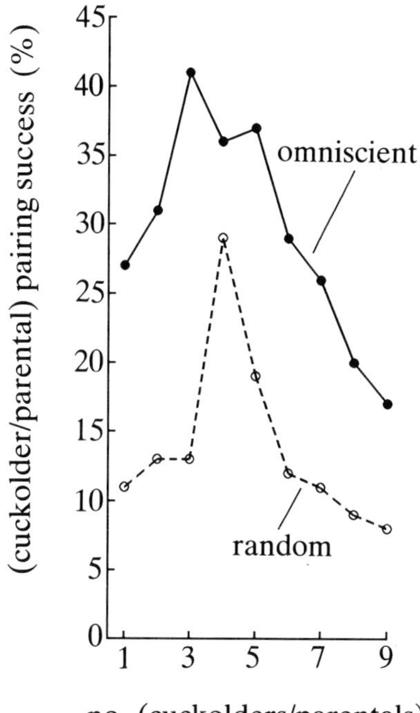

Figure 2. The relative pairing success of cuckolder and parental males ($\bar{W}c/\bar{W}p$) across a range of frequencies (Nc/Np) in the population. Two curves have been calculated based on assumed cuckolder distributions among the bluegill colonies in figure 1: omniscient and random (see text for details).

(*c*) ***Frequency-dependent pairing success***

The theoretical calculations of the random and omniscient cuckolder distributions among the four colonies show that the relative average pairing success of cuckolder and parental males depends upon their relative frequencies in the population (figure 2, example calculations in Appendix 1). With an omniscient distribution, and an equal frequency of cuckolder and parental males in the population ($Nc/Np = 1$), cuckolders will obtain about 27% as many pairings as parental males. This increases to 41% when cuckolders outnumber parentals 3:1 ($Nc/Np = 3$), but declines to 17% at 9:1 ($Nc/Np = 9$).

As expected, cuckolders are less successful with a random distribution among colonies. With an equal frequency of cuckolder and parental males in the population ($Nc/Np = 1$), cuckolders obtain only 11% of the success of parentals. This increases to 29% the success when cuckolders outnumber parentals 4:1 ($Nc/Np = 4$), but declines to 8% at 9:1 ($Nc/Np = 9$). Therefore both distributions result in negative frequency-dependent success once cuckolder frequency becomes high in the population.

4. DISCUSSION

The experimental manipulations of cuckolder density within bluegill colonies of Lake Opinicon have shown that cuckolder pairing success is density dependent, but that there is a unique density at each colony which maximizes individual success. The population is therefore heterogeneous in the relation between cuckolder density and success. Consequently, no single colony can provide an adequate test of frequency-dependent selection between the cuckolder and parental strategies within the population.

The occurrence of unique densities where cuckolder success peaks is largely due to ecological differences in cover among colonies. Cover is important because it determines how closely cuckolders can approach a nest. Close proximity improves a cuckolder's ability to monitor spawning activity, and also aids his ability to pair with the female by reducing the length of his exposure to the parental male's aggressive defence. In colonies with abundant cover, additional cuckolders at the nest site increase competition, and therefore average cuckolder success is negatively density dependent immediately. But when cover is sparse, cuckolders are unable to stay close to the nest. Additional cuckolders then make a positive contribution to individual mating success by indirectly creating opportunities for each other to intrude through distraction of the parental male. Thus individual cuckolder success in colonies with sparse cover is positively density dependent at first. Because increasing cuckolder density eventually creates more competition than distraction, individual success then becomes negatively density dependent. Each colony thus has an optimum cuckolder density, dependent on cover, at which cuckolder pairing success is maximized.

The colony-specific optimums in cuckolder density will be a key factor in how natural selection favours cuckolders to distribute themselves among colonies. However, there are several constraints that may also influence cuckolder distribution, preventing the fitness of the cuckolder strategy from being maximal. For example, predators hunting in deeper colonies (Gross 1982) may favour the crowding of cuckolders into shallower colonies with more cover. A trade-off between survival and reproductive success will result in lower individual pairing success owing to increased competition (for example, see Lima & Dill 1990). Another constraint on distribution is the information available to cuckolders about the relative advantage of different colonies. Since parental males use traditional nesting sites, cuckolders could be aware of these locations and the ecological situations there. Cuckolders could also know their own abundance and density as cuckolders often travel in schools. A third factor that may influence distribution is despotic behaviour by fellow cuckolders. Once the optimum density was reached, cuckolders were observed to become increasingly aggressive towards each other; larger individuals sometimes drove smaller individuals from nest sites. This behaviour may displace some cuckolders to alternative colonies where individual success is lower. Such a despotic distribution (Fretwell & Lucas 1970) would not only increase the variance among individuals, but it could also decrease the mean fitness of the strategy.

Cuckolders will therefore distribute themselves among colonies based on the above constraints. How well they do this to maximize their own success ultimately determines the cumulative success of the cuckolder strategy itself and thus its fitness relative to the parental strategy.

Our theoretical model of an omniscient distribution allowed cuckolders to distribute themselves according to the unique density-dependent relations within colonies, free of any constraints. In this case, the marginal value of the colonies is reduced equally and the fitness of the cuckolder strategy is as high as it can possibly be. In our random distribution, cuckolders followed the dispersion of parental males regardless of ecological differences among colonies. Thus cuckolders were not allowed to adjust their densities to maximize their mating success. Using the omniscient distribution, on average the cuckolder strategy achieved 15% more pairing success relative to the parental strategy than did the random distribution (figure 2). This shows that the distribution pattern can have a significant effect on relative fitnesses of the strategies. The analysis also showed, however, that at high cuckolder frequencies, the cuckolder's strategy's fitness declines relative to the parental strategy's fitness in *both* distributions. This shows that as their frequency increases, there is greater within-strategy competition among cuckolders than between the cuckolder and parental strategies themselves. Moreover, the within-strategy competition cannot be beaten by an omniscient distribution. Therefore, parental males are always able to maintain sufficient paternity to cause average cuckolder success to decline as cuckolder density increases.

What prevents parental males from maintaining their paternity completely? The explanation must involve the time constraint under which females operate during spawning. Female spawning behaviour has evolved under selection for synchronous spawning to minimize brood predation (Gross & MacMillan 1981). Spawning in a colony occurs in a short time span, and females do not tolerate much limitation of their dipping rate by aggression from a parental male. Parentals are thus constrained in their ability to control the female's spawning and must therefore trade off loss of fertilizations to cuckolders with fewer eggs spawned into their nest. They can forfeit fertilization of some dips to cuckolders while allowing the female to spawn rapidly, or they can slow the female's dipping rate for better control of the fertilizations but receive fewer eggs and risk female desertion. When parental males cannot maintain this compromise at extreme cuckolder densities, it becomes more favourable for the parental male to abort a spawning by driving the female from his nest rather than forfeiting all fertilizations of her dips to cuckolders.

Therefore, the time constraint on female spawning behaviour may prevent parental males from monopolizing all the spawnings, evolutionarily allowing cuckolders to exist. But as cuckolder frequency increases, the number of parental males must decrease and so too the number of alternative nests available to females. As alternative nests become less available, female choosiness must, on average, decrease. This would allow the parental male to control more of the spawning. It is this linkage among the constraints and tactics of female, parental, and cuckolder bluegill that permits existence of an alternative male reproductive strategy.

The life-history models of Gross & Charnov (1980) and Gross (1982) predict that the ESSt frequency of cuckolder and parental males will exist where their lifetime fitnesses are equal. This occurs when the proportion of males in the population that are cuckolders, q, is equal to the proportion of eggs in the population that cuckolders fertilize, h (i.e. $q = h$). In Lake Opinicon, the age class at which most cuckolders first mature is age two, and q is about 11–31% (95% binomial confidence interval; Gross 1982). At this frequency, the earlier sexual maturity of cuckolders and decreased survivorship of males in the parental life history results in a ratio of about six mature cuckolders to one mature parental male in the breeding population. Referring to figure 2, this 6:1 ratio produces a mating success of cuckolders remarkably close to the value of h that results in equal lifetime fitness. For example, h ranges from 11% to 23% in the random and omniscient distributions, respectively, when cuckolders outnumber parental males six to one (to calculate h, the y-axis is converted from relative pairing success to percent of total pairings; e.g. h for the omniscient distribution is calculated by converting the 29% on the y-axis to $29/[100+29] = 23\% = h$). Because h increases relative to q at lower cuckolder frequencies, and h decreases relative to q at higher cuckolder frequencies, the frequency-dependent sexual selection in Lake Opinicon can hold the two strategies at an ESSt balance point.

I thank the Society for the Study of Evolution for supporting this work through the 1981 Theodosius Dobzhansky Prize. My research on bluegill has continued through NSERC of Canada Operating Grants. I thank E. L. Charnov, S. Rohwer and N. L. Gerrish for research discussions, E. P. van den Berghe for field assistance, R. M. Coleman and especially N. L. Gerrish and L. Dueck for improving the manuscript, R. Robertson and F. Phelan for facilities at the Queen's University Biological Station, and P. Harvey and L. Partridge for the opportunity to present this work in London.

REFERENCES

Charnov, E. L. 1982 *The theory of sex allocation.* Princeton University Press.

Darling, F. F. 1937 *A herd of red deer.* Oxford University Press.

Fretwell, S. D. & Lucas, H. L. 1970 On territorial behaviour and other factors influencing habitat distribution in birds. *Acta biotheor.* **19**, 16–36.

Gadgil, M. 1972 Male dimorphism as a consequence of sexual selection. *Am. Nat.* **106**, 574–580.

Gross, M. R. 1979 Cuckoldry in sunfishes (*Lepomis*: Centrarchidae). *Can. J. Zool.* **57**, 1507–1509.

Gross, M. R. 1980 Sexual selection and the evolution of reproductive strategies in sunfishes (*Lepomis*: Centrarchidae). Ph.D. thesis. University of Utah, Salt Lake City. University Microfilms Int., No. 8017132, Ann Arbor, Michigan.

Gross, M. R. 1982 Sneakers, satellites and parentals: polymorphic mating strategies in North American sunfishes. *Z. Tierpsychol.* **60**, 1–26.

Gross, M. R. 1984 Sunfish, salmon and the evolution of alternative reproductive strategies and tactics in fishes. In *Fish reproduction: strategies and tactics* (ed. G. Potts & R. Wootten), pp. 55–75. London: Academic Press.

Gross, M. R. 1985 Disruptive selection for alternative life histories in salmon. *Nature, Lond.* **313**, 47–48.

Gross, M. R. 1991 The evolution of alternative tactics and strategies. *Behav. Ecol.* (In the press.)

Gross, M. R. & Charnov, E. L. 1980 Alternative male life histories in bluegill sunfish. *Proc. natn. Acad. Sci. U.S.A.* **77**, 6937–6940.

Gross, M. R. & MacMillan, A. M. 1981 Predation and the evolution of colonial nesting in bluegill sunfish (*Lepomis macrochirus*). *Behav. Ecol. Sociobiol.* **8**, 163–174.

Hutchings, J. A. & Myers, R. A. 1988 Mating success of alternative maturation phenotypes in male Atlantic salmon, *Salmo salar. Oecologia* **75**, 169–174.

Jones, J. W. 1959 *The salmon.* London: Collins.

Kindler, P. M., Philipp, D. P., Gross, M. R. & Bahr, J. M. 1989 Serum 11-ketotestosterone and testosterone concentrations associated with reproduction in male bluegill (*Lepomis macrochirus*: Centrarchidae). *Gen. comp. Endocrinol.* **75**, 446–453.

Knoppien, P. 1985 Rare male mating advantage: a review. *Biol. Rev.* **60**, 81–117.

Lank, D. B. & Smith, C. M. 1987 Conditional lekking in ruff (*Philomachus pugnax*). *Behav. Ecol. Sociobiol.* **20**, 137–145.

Lima, S. L. & Dill, L. M. 1990 Behavioral decisions made under the risk of predation: a review and prospectus. *Can. J. Zool.*, **68**, 619–640.

Maekawa, K. & Onozato, H. 1986 Reproductive tactics and fertilization success of mature male Miyabe charr, *Salvelinus malma miyabei. Env. Biol. Fish* **15**, 119–129.

Maynard Smith, J. 1982 *Evolution and the theory of games.* Cambridge University Press.

Parker, G. A. 1984 Evolutionarily stable strategies. In *Behavioural ecology: an evolutionary approach* (ed. by J. R. Krebs & N. B. Davies), pp. 30–61. Sunderland, Massachusetts: Sinauer Associates.

Partridge, L. 1988 The rare-male effect: what is its evolutionary significance? *Phil. Trans. R. Soc. Lond.* B **319**, 525–539.

Partridge, L. & Hill, W. G. 1984 Mechanisms for frequency-dependent mating success. *Biol. J. Linn. Soc.* **23**, 113–132.

van Rhijn, J. G. 1983 On the maintenance and origin of the alternative strategies in the ruff *Philomachus pugnax. Ibis* **125**, 482–498.

Warner, R. R. 1984 Mating behavior and hermaphroditism in coral reef fishes. *Am. Scient.* **72**, 128–136.

APPENDIX 1

Both the random and omniscient cuckolder distributions are calculated from $Np = 62$ (the total number of parental males in the Pen Bay population). The parental distribution, Pp_i, is the empirical distribution observed among the four colonies (table 1). The number of females spawning in colonies is assumed to follow the parental distribution. To calculate the relative success of the cuckolder and parental strategies from the random distribution, let $Pc_i = Pp_i$. For example, with equal numbers of cuckolder (62) and parental (62) males in the population, $Nc/Np = 1$; Pc_A (proportion of cuckolders at colony A) $= Pp_A = 0.194 = 12$ males; $Pc_B = Pp_B = 0.226 = 14$ males; $Pc_C = Pp_C = 0.194 = 12$ males and

$Pc_D = Pp_D = 0.387 = 24$ males. In this case there is one cuckolder at each nest of a parental male. The empirically measured pairing success per cuckolder in figure 1 shows that $\bar{S}c_A = 16.6$, $\bar{S}c_B = 17.7$, $\bar{S}c_C = 2.9$ and $\bar{S}c_D = 5.3$. Entering these values into equation (1) gives 609.0, equation (2) is 5592.4 and equation (3) is 0.109. Thus with $Nc/Np = 1$, a cuckolder male in a random distribution averages 11% the success of a parental male in the population. Similar calculations were made for all $Nc/Np \leqslant 9$, giving the data plotted in figure 2.

The relative success of the cuckolder strategy in the omniscient distribution is calculated by varying Pc_i independently of Pp_i, and maximizing equation (1) by simulation. For example, with $Nc/Np = 1$, the distribution of cuckolders among colonies that will maximize equation (1) is: $Pc_A = 0.032$ (1 cuckolder at 2 nests); $Pc_B = 0.452$ (2 cuckolders per parental); $Pc_C = 0.520$ (4 cuckolders at 8 nests) and $Pc_D = 0$ (no cuckolders go to colony D). It follows that equation (1) $= 1326.8$, equation (2) $= 4873.2$ and equation (3) $= 0.272$. Thus with $Nc/Np = 1$, a cuckolder male in the omniscient distribution averages 27% the success of parental males in the population (figure 2). For a second example, consider $Nc/Np = 3$ (3 cuckolders per parental male in the population). The cuckolder distribution maximizing equation (1) is now: $Pc_A = 0.065$ (1 cuckolder per parental); $Pc_B = 0.151$ (2 cuckolders per parental); $Pc_C = 0.258$ (4 cuckolders per parental) and $Pc_D = 0.527$ (4 cuckolders per parental at 22 nests and 5 cuckolders per parental at 2 nests). Equation (3) $= 0.406$. Therefore, with cuckolder males outnumbering parental males 3:1, a cuckolder male in an omniscient distribution averages 41% the success of a parental male in the population. Similar calculations were made for all $Nc/Np \leqslant 9$.

Clutch size, fecundity and parent–offspring conflict

H. C. J. GODFRAY[1] AND G. A. PARKER[2]

[1] *Department of Biology & NERC Centre for Population Biology, Imperial College at Silwood Park, Ascot, Berkshire SL5 7PY, U.K.*
[2] *Department of Environmental and Evolutionary Biology, University of Liverpool, Liverpool L69 3BX, U.K.*

SUMMARY

Selection often acts in different ways on genes expressed in parents and offspring leading to parent–offspring conflict. The effect of parent–offspring conflict on the evolution of reproductive strategies is explored. Models are constructed using kin-selection techniques and it is argued that these are frequently more useful than techniques from classical population genetics. Parent and offspring optima are compared in models of (1) the trade-off between the number and size of offspring, (2) clutch size and (3) the evolution of reproductive effort with age structure. Parent–offspring conflict over clutch size is examined in more detail. Models of sibling competition are reviewed and it is suggested that the reduction in parental fitness caused by sibling competition may lead to selection on clutch size. The possibility that the parent may be selected to produce a hierarchy of sizes of young in order to reduce sibling conflict is investigated. The preliminary results give little support for this hypothesis. An extreme form of sibling conflict, siblicide, is also discussed. In some cases, the kin-selection approach fails in the analysis of siblicide and classical population genetic models are required. The paper concludes that parent–offspring conflict is a potentially significant, and often overlooked, factor influencing the evolution of reproductive strategies.

Birds in their little nests agree
And 'tis a shameful sight,
When children of one family
Fall out, and chide, and fight
Isaac Watts, *Love between Brothers & Sisters*, 1721.

Birds in their little nests agree
With Chinamen, but not with me.
Hilaire Belloc, *On Food.*

1. INTRODUCTION

It is a truism to state that family life is seldom unmoderated bliss. Yet biologists analysing life-history strategies have classically assumed harmony of purpose between family members. An obvious case concerns Lack's (1947) pioneering theory of clutch size. Lack assumed that selection would favour the behaviour that generates the maximum number of surviving offspring from a given clutch. This can be optimal for a single parent, but it need not be so for two parents, nor for the offspring.

Although field workers had long and variously described the sometimes quite dramatic manifestations of sib competition, theorists in general ignored such facts when constructing models. The major theoretical development came from Hamilton (1964). Although he was primarily concerned with the problem of altruism between siblings (and other relatives), Hamilton's now-famous rule applies in reverse, to describe the limits of sibling-selfishness. Hamilton also realized that in an evolutionary sense, parent and offspring interests need not coincide. In a remarkable paper, Trivers (1974) used Hamilton's rule to formulate a theory of parent–offspring conflict for sexually reproducing species: offspring should demand more parental investment than serves parental interests.

A backlash against the idea of parent–offspring conflict came swiftly from Alexander (1974); he argued that a gene that causes the summed fitness of a brood to be reduced would be quickly eliminated, in effect because offspring that conflict become parents that produce conflicting offspring. The parent should therefore always win. However, this argument treats parental fitness as paramount and, as Dawkins (1976) pointed out, rephrasing the argument giving primacy to offspring fitness leads to the conclusion that the offspring always wins. The essential consideration is whether a gene that causes offspring to conflict can spread against its alternative allele (for not conflicting). Whereas Alexander's argument is correct for asexuality, the translation of parental to offspring characteristics is imperfect in sexual species; carriers of a mutant conflictor gene can therefore profit at the

Phil. Trans. R. Soc. Lond. B (1991) **332**, 67–79
Printed in Great Britain

expense of their non-conflicting sibs. Alexander's paper had the useful effect of stimulating a series of studies based on explicit population genetics (Stamps *et al.* 1978; Parker & Macnair 1978, 1979; Macnair & Parker 1978, 1979; Charnov 1982; Parker 1985). These showed unequivocally that such 'conflictor genes' could spread. Alexander (1974) also argued that parent–offspring conflict is implausible as the parent will always be in a position to impose its optimum.

Despite the controversy, parent–offspring conflict has not been a popular subject with empiricists, perhaps because tests of parent–offspring conflict theory often do not easily lend themselves to the dominant manipulative–experimental research programme of field behavioural ecology. One notable exception concerns parent–offspring conflict over sex ratio in social Hymenoptera (Trivers 1974). Trivers & Hare (1976) produced evidence – from the ratio of investment in males and females – that workers (offspring) control sex ratio in defiance of the parental optimum. This conclusion was quickly challenged by Alexander & Sherman (1977) and is still the subject of active research (Nonacs 1986; Boomsma 1989; Boomsma & Grafen 1990). Much of the other evidence for parent–offspring conflict comes from qualitative and even anecdotal evidence on behavioural interactions such as conflict over the time of weaning and over the amount of food provisioning during parental care, which often does not greatly extend Trivers' (1974) own comments (see Clutton-Brock (1991) for a recent review).

Theoretical studies of parent–offspring conflict have been concentrated in a few quite well circumscribed areas. Perhaps the most attention has been paid to competition between nestling birds (see below) though other active have included conflict over the sex ratio (Hamilton 1967; Trivers & Hare 1976; Charnov 1982), over infanticide (O'Connor 1978; Hausfater *et al.* 1982), and over the allocation of resources to different reproductive tissues in plants (Haig 1987). However, parent–offspring conflict is potentially significant whenever offspring and parents are both able to influence the realization of a life-history trait. The primary aim of this paper is to explore the consequences of parent–offspring conflict for some of the classic problems of life-history theory. The next section discusses techniques for the analysis of models of parent–offspring conflict and then the following three sections treat (i) the trade-off between number and size of offspring, (ii) clutch size and sibling conflict, and (iii) age-specific reproductive effort. We dwell longer on clutch size and sibling conflict as this area has received much theoretical attention. A subsidiary aim of this paper is to provide a non-mathematical review and summary of this work.

2. A NOTE ON METHODOLOGY

Initial concern about the population genetic underpinning of the subject led to the proleptic incorporation of explicit genetics in many models of parent–offpsring conflict (Parker & Macnair 1978, 1979; Macnair & Parker 1978, 1979; Parker 1985; Stamps *et al.* 1978; Feldman & Eshel 1982; Harper 1986). In comparison with phenotypic models, explicitly genetic models are more rigorous and allow the study of evolutionary dynamics, as well as end points. However, genetic models are normally complicated to analyse and the added realism of explicit genetics is rather spurious as seldom if ever is the actual genetic basis of the trait known. A related problem is the uncertainty about the degree to which the results of the analysis depend on the specific underlying genetic assumptions.

An alternative to genetic models is the use of phenotypic models. As parent–offspring conflict concerns interactions between relatives, models must be couched in terms of inclusive fitness. Hamilton's rule provides the easiest technique for studying parent–offspring conflict (Trivers 1974). The rule states that a trait will spread if the sum of the weighted changes in the fitness of self and relatives brought about by the trait are positive. The weights used in comparing fitnesses are the coefficients of relatedness of self to relatives. Hamilton's rule provides conditions for the spread of a trait and, for non-competitive problems, also defines the evolutionary equilibrium. Thus, if an older chick completely determines the division of food between itself and a younger sibling, the optimum resource division for the older chick is found by a simple application of Hamilton's rule.

Many problems in parent–offspring conflict concern competition between different family members necessitating an evolutionarily stable strategy (ESS) approach. The standard technique for the analysis of these problems is to assume single-locus genetics. However, simpler and more general results can be obtained using a modified form of Hamilton's rule (Godfray & Parker 1991). Assume that the population is at the ESS and that a mutant arises with a deviant behaviour. At the ESS the costs and benefits of an arbitrarily small change of behaviour should be equal. The ESS can thus be found by assessing the costs and benefits of deviant behaviour using Hamilton's rule, and then taking the limit as the magnitude of the deviance approaches zero. This approach can be viewed as a simple extension of that adopted by Trivers (1974) as it uses kin-selection methodology rather than classical population genetics. The 'Marginal' Hamilton's rule works for continuous behaviour strategies with weak selection and additive phenotypic effects.

The assumptions of weak selection and additivity will be violated in some cases of parent–offspring conflict. For example, the effects of siblicide and infanticide are obviously non-additive. In these circumstances, it can be essential to use explicit population genetic models.

The study of parent–offspring conflict can be divided into two stages. Initially the optimum strategies for both the parent and young are calculated and compared. Any discrepancy between the two optima defines the battleground within which the conflict is waged. The second stage is to study the resolution of the conflict and this tends to involve a second series of assumptions. In the simplest case, the offspring or the parent are assumed to have complete control over the

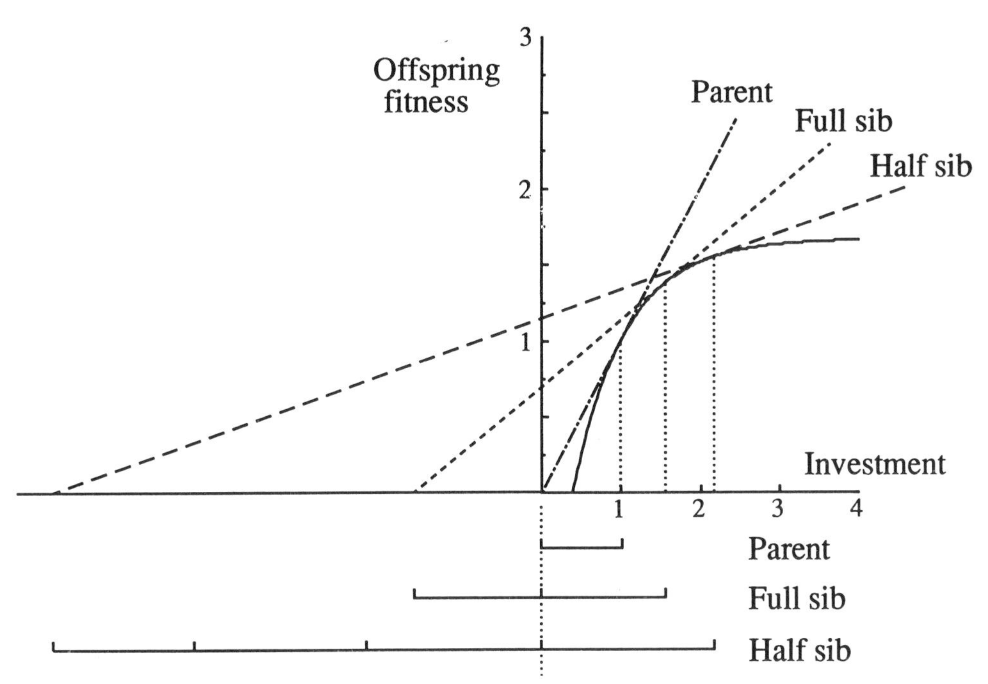

Figure 1. Graphical illustration of the Smith & Fretwell model and its extension to calculating the offspring optima. The solid curve describes offspring fitness as a function of investment. The parental optimum is defined as the point where a tangent rooted at the origin touches the curve. The optimum for full and half sibs can also be defined as the point where a rooted tangent touches the curve. In the case full sibs, the tangent is rooted at a distance to the left of the origin equal to the optimum investment. In the case of half sibs, the root of the tangent is at a distance three times the optimum investment to the left of the origin.

realization of the trait in question and thus one side unconditionally 'wins'; we shall give examples of both parental and offspring triumphs below. Alternatively, both sides may have partial control over the realization of the trait and the predicted strategy is the stimultaneous parent–offspring ESS (which can be calculated using the marginal Hamilton's rule). In this paper we are chiefly concerned with the first of these stages, although in some cases we discuss the likely resolution of the conflict.

3. INVESTMENT PER OFFSPRING

Consider a parent with a fixed amount of resource that it shares between an indeterminate number of offspring. The evolution of this decision will be influenced by the trade-off between the number and quality of young. As a specific example of this trade-off, consider an invertebrate whose only investment in its young is the resources (e.g. yolk) that make up the egg. The animal will experience a trade-off between the size and the number of the eggs it produces (Smith & Fretwell 1974; Parker & Begon 1986; Lloyd 1987). Alternatively, consider an animal, perhaps a mammal, that has a finite amount of resources that it can allocate to a series of successive young (we shall discuss more sophisticated models below). The animal will again experience a trade-off, this time between the number of young it produces, and the amount of resources it allocates to individual offspring (Parker & Macnair 1978; Macnair & Parker 1978; Winkler & Wallin 1987).

Smith & Fretwell (1974) first showed that the optimal investment per offspring could be found using a simple graphical model (figure 1). In words, the optimal investment occurs at the point when the marginal gain in offspring fitness, brought about by an arbitrarily small increase in investment, is exactly counterbalanced by the reduction in the number of offspring consequent on this resource redistribution. The parent balances the benefits of fitter offspring against the reduction in number of offspring: note, as the parent is equally related to all her offspring, she weights all offspring identically.

Smith & Fretwell calculated the optimum investment per offspring from the point of view of the parent. We can now investigate the scope for parent–offspring conflict by calculating the optimum investment per offspring from the point of view of the offspring. Consider the case where all investment is made by a single parent (Parker (1985) explains why assumptions about whether one or two parents invest in the young are important). Again, the optimum will be set by the balance between fitter offspring and the number of offspring. However, whereas the parent valued all offspring identically, any particular offspring devalues other offspring by the coefficient of relatedness between siblings. At the offspring ESS, the marginal gain in offspring fitness with resources is thus less than at the parental ESS and hence investment per offspring is larger and parental fecundity is smaller (Parker & Macnair 1978; Macnair & Parker 1978; Lazarus & Inglis 1986, Clutton-Brock & Godfray 1991). This result can also be shown graphically (figure 1).

Thus, if offspring have their way, parental investment is higher and overall parental fecundity is lower. To quantify this statement, it is necessary to specify a particular form for the relation between

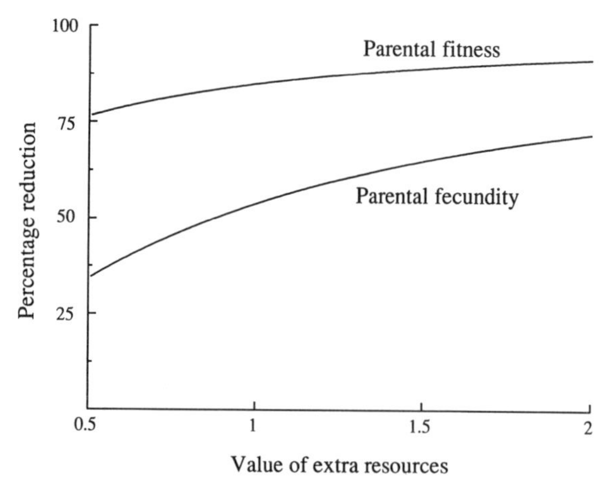

Figure 2. Parental fitness and fecundity (expressed as a percentage of the maximum) when the offspring determine the trade-off between the number and fitness of young. These results were obtained using a specific relation between offspring fitness and resources described in Parker & Macnair (1978). The rate at which offspring fitness increases with additional resources beyond the parental optimum is determined by a single parameter which is plotted on the x axis. Low values imply a rapid gain of fitness and hence greater benefits for conflict.

offspring fitness and parental investment (i.e. the shape of the curve in figure 1). The most important feature of this relation is the rate at which offspring fitness increases with extra resources beyond the parental optimum: more conflict is expected when the profits are greater. Figure 2 illustrates the decline that can occur in parental fecundity. Depending on the shape of the offspring fitness curve, quite considerable reductions can be predicted. Note that although parental fecundity can drop quite spectacularly, the reduction in parental fitness is less impressive. Although the parent has fewer offspring, at least they are fitter, and this acts in partial recompense.

How likely are offspring to be able to influence resource allocation by parents? It seems unlikely that a single ovum could prevail upon the mother to increase expenditure beyond her optimum, although it is perhaps unwise to underestimate potential 'biochemical conflict' between mother and young. However, the Smith & Fretwell model also works as a model, admittedly only very approximate, of mammals and birds with sequential offspring. If birds and mammals monitor their young to assess resource needs (see also below) there is great potential for guile and misinformation by the young. Thus this framework offers a rough guide to the potential life-history consequences of such manifestations of parent–offspring conflict as weaning conflict.

4. CLUTCH SIZE

We now turn to the evolution of clutch size. As the number of individuals in a clutch increases, the average fitness of individual offspring decreases. This decline in offspring fitness is normally caused by competition between members of the clutch for limiting resources (Lack 1947) but may be the result of many other factors. For example, the risk of predation may increase with clutch size if predators are attracted by the total size or noise output of a brood of animals (Skutch 1949). Clutch size will also be influenced by trade-offs between the size of the clutch and the future reproductive success of the parent (Charnov & Krebs 1974, Godfray 1987 *b*). However, in this section we will assume that clutch size is determined solely by the first trade-off, between number and fitness of offspring. In these circumstances, a parent will be selected to produce the clutch size that maximizes the fitness returns for that clutch. In the simplest case, when all offspring have identical fitness, this reduces to the problem of maximizing the product of the number of offspring and average fitness. It has become customary to refer to this result as the 'Lack clutch size' as it is a general vision of the explanation given by Lack (1947) for avian clutch sizes.

Consider a bird that a produces a Lack clutch of eggs. Ideally, the parent would prefer each of its offspring to take its fair share of resources and certainly not to engage in any form of 'intrabrood' competition that may reduce their fitness. In the first part of this section we examine how competition between offspring in a brood may frustrate the parental desideratum. We suppose that the young can garner a greater than fair share of resources by increasing their intensity of competition, for example by begging more loudly than their siblings. We show that when there are fitness penalties attached to increased competition, the ESS level of competition among the young can lead to a marked decrease in parental fitness. In the second part of this section we show that competition between siblings can lead to selection on the parent to reduce its clutch size. Although we have used sibling birds as an illustrative example, we have so far ignored a major feature of bird biology, that nestlings typically form a hierarchy of sizes and competitive abilities. In the third part of this section we examine how this assumption affects our conclusions and ask whether the chick hierarchy has arisen as an adaptation to reduce harmful sibling conflict. In the final part of the section, we examine perhaps the most extreme form of sibling conflict, siblicide: the deliberate destruction of one sibling by another.

(*a*) *Sibling conflict and the fitness of the clutch*

The evolution of sibling conflict has most often been studied in the context of competition between sibling birds and the evolution of begging. We shall continue in this tradition though discuss other examples later. Our account here attempts to synthesize the theoretical studies of Stamps *et al.* (1978), Macnair & Parker (1979), Parker & Macnair (1979), Lazarus & Inglis (1986), Harper (1986) and Godfray & Parker (1991). Note, we are concerned here with what Parker & Macnair (1978) called intrabrood competition: competition between members of the same brood. This is in contrast to interbrood competition, competition for resources between broods, obviously mediated through the parents.

To study the evolution of begging it is necessary to state how increased begging translates into increased resources. In fact, all workers have adopted a mechanism that might be called *mean matching* where the share of resources obtained by an individual is influenced by the *relative* intensity of its begging in comparison with the average for the brood (note Harper (1986) is incorrect in saying that the absolute, rather than the relative, begging intensity is important in the models of Parker & Macnair). The mechanism of mean matching implicitly incorporates two biologically important features. First, it implies that increased begging is more efficient in large broods. The reason for this is that in small broods a change in behaviour by one individual has a marked effect on the mean and so lessens the contrast between that individual and the brood average. We have referred to this as the dilution effect. Second, the efficiency of increased begging declines as the background level of begging increases. Begging at an extra decibel is more impressive if the background begging is one decibel than if it is ten decibels. The analysis here is restricted to mean matching though other mechanisms are possible. A possible alternative mechanism suggested to us by M. R. Macnair is that the parent compares the begging intensity of an individual with the begging intensity of the rest of the brood.

As we have assumed that all chicks in a brood are identical, at the ESS all chicks will beg at the same rate. Suppose there are no costs to begging, the ESS can never be stable as small unilateral increases in begging will always be rewarded and never punished. In reality, begging will, at least at some level, incur costs. The manner in which the costs are distributed across the brood members is crucial in predicting the ESS level of begging. Godfray & Parker (1991) distinguish three exemplar forms of cost:

1. Individual costs: all costs of begging are experienced by the individual. This might occur when the costs of begging are exclusively metabolic.
2. Shared costs: the costs experienced by an individual are proportional to the average begging in the brood. Some predators may respond to the mean level of begging emanating from a brood.
3. Summed costs: the costs experienced by an individual are proportional to the sum of all individual's begging. Some predators may respond to the total level of begging emanating from a brood.

(The series of papers by Parker & Macnair used cost forms (1) and (2) while Harper (1986) used cost forms (1) and (3).) The fitness of the young is thus influenced by two components, their share of the limiting resource, and the costs of begging. Different workers have combined these two components in different ways to obtain overall fitness (for example multiplicatively or additively). However, the qualitative results of the model seem robust to this detail.

The ESS level of begging can be found by using the marginal form of Hamilton's rule. At the ESS, the personal benefits of a microscopic increase in resources are exactly balanced by the costs to relatives of reduced resource share and the costs to self and, in the case of most forms (2) and (3), to relatives of increased begging. Of course, costs and benefits to relatives are weighted by the coefficient of relatedness.

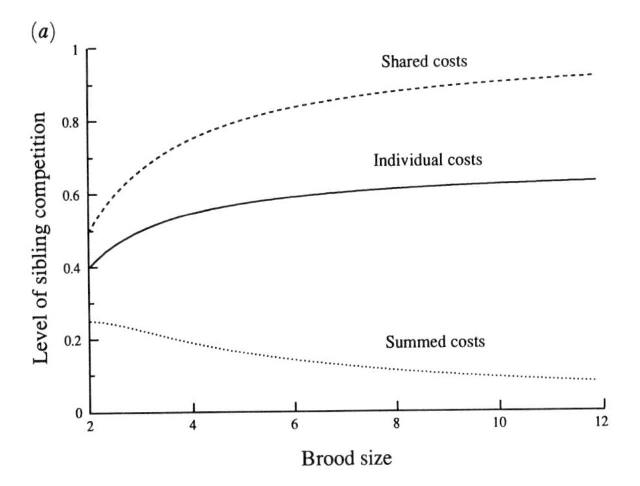

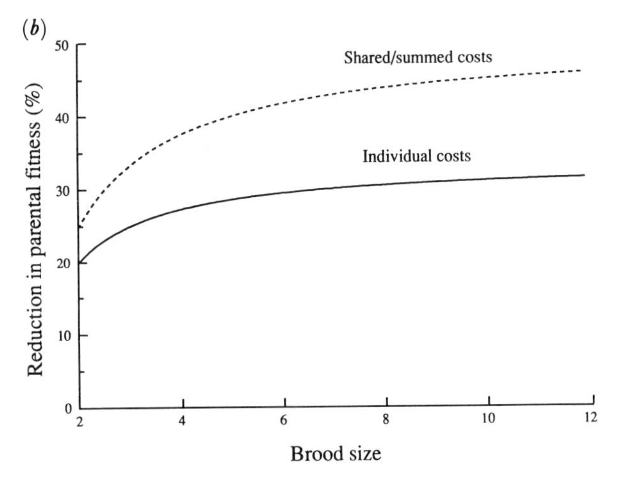

Figure 3. (*a*). The relation between the ESS levels of begging and brood size for three different forms of costs. (*b*) The reduction in parental fitness as a consequence of ESS levels of begging at different clutch sizes and with different forms of costs. (After Godfray & Parker 1991.)

Some typical predictions of ESS begging levels are shown in figure 3 (after Godfray & Parker 1991). The graphs show the level of sibling competition and the reduction in parental fitness as a function of brood size for the three forms of costs described above. Levels of begging are higher for shared costs than for individual costs (Macnair & Parker 1978) but higher for individual costs than summed costs (Harper 1986). Begging increases with brood size in the case of individual costs (Harper 1986) and shared costs (Godfray & Parker 1991) but the reverse occurs with summed costs (Harper 1986). The reduction in parental fitness mirrors these changes except in the case of summed cost which, perhaps counter-intuitively, are the same as those for shared costs (Godfray & Parker 1991). To draw figure 3, we assumed the coefficient of relatedness among siblings was $\frac{1}{2}$. The models predict begging intensity to be inversely related to the coefficient of relatedness (Macnair & Parker 1979).

Lazarus & Inglis (1986) also discuss the relation between parent–offspring conflict and clutch size. Unlike the other studies discussed in this section, they assumed that the total investment in a brood was subject to parent–offspring conflict while intrabrood conflict was either absent or represented by a static dependence of offspring fitness on clutch size. Further work is needed to examine models incorporating both intrabrood and interbrood conflict.

To understand these results, consider the case of individual costs. Because of the dilution effect (see above), the net benefits of relatively louder begging increase with brood size (rapidly at first, but then approaching a plateau) while the costs to the individual remain constant. The trade-off between these two factors gives rise to louder ESS level of begging in large clutches. The dilution effect also occurs in the case of summed and shared costs. However, in these cases the cost term is also influenced by clutch size. With shared costs, the negative consequences of increased begging are divided between all members of the brood and thus become less important as clutch size increases. The reverse happens with summed costs where the harmful effects of individual begging are magnified by clutch size. The change in costs with brood size interacts with the dilution effect to accentuate the increase in begging with brood size in the case of shared costs but to reverse the trend in the case of summed costs. Finally, the reduction in parental fitness is proportional to begging levels in the case of individual and shared costs. In the case of summed costs, the reduction is proportional to the product of begging intensity and clutch size. In the particular model analysed here, this results in the same reduction of parental fitness for summed and shared costs.

To summarize this section, competition between siblings can lead to marked reductions in the fitness of clutches of animals. The exact reduction will depend both on clutch size and on how the costs of competition are spread amongst the brood. Although we have exclusively talked about begging in birds, we could have easily analysed other forms of sibling competition. For example, Godfray & Parker (1991) discuss a model of feeding in gregarious caterpillars. Instead of begging, caterpillars are allowed to alter their rate of food intake. Faster feeding gives an individual a competitive edge but leads to costs, either to the individual (metabolic costs or costs associated with reduction in the efficiency of assimilation) or to the brood (increased consumption of plant material may reduce plant growth and thus the total amount of resource available to the brood).

There is thus a wide battleground in which parent–offspring conflict can occur. Who wins the conflict? Alexander (1974) argued that the parent always wins as it determines resource share. This argument has some force for birds and other species where resources are doled out by the parent, but it does not apply to the gregarious caterpillars discussed above where resource share is determined solely by competition between siblings. In the latter case, the offspring always wins. Dawkins (1976) argued that even in species with parental control of resource share, parent–offspring conflict can still occur as efficient resource distribution requires the parent to monitor offspring needs and this allows the offspring to misrepresent their requirements. In an elegant model, Harper (1986) showed that parents will be selected to adjust their resource share to begging level, as long as some of the variance in begging reflects true need. Parker & Macnair (1979) also argue that the strategy of 'ignore solicitation' may be unstable when there are costs to ignoring offspring begging. The same authors modelled the evolution of parental retaliation by assuming that the extent to which the parent responds to increased begging is controlled by natural selection. They predicted that the joint ESS will be intermediate between the parent and offspring optima.

To test the models discussed in this section, two types of study are needed. First, experimental studies of the relation between begging level and resource share, and of the distribution of costs among individual broods are required. There is already some information on this subject in the ornithological literature (see, for example, Ryden & Bengtsson 1980; Bengtsson & Ryden 1981) but there has yet been no systematic attempt to verify the assumptions of begging models. The second type of study needed is direct tests of the model predictions. The relation between begging levels and clutch size can be tested both intraspecifically and interspecifically. Harper (1986) has discussed interspecific patterns that support the predictions of begging models but a more formal comparative study is required. The predictions of the models might also be tested by relating begging levels to the type and severity of the costs associated with begging. For example, birds nesting in holes suffer lower predation rates than those with open nests (Ricklefs 1969) and tend to beg more noisily (Harper 1986). Finally, the dependence of begging levels on relatedness within broods may offer opportunities to test theory. Harper (1986) suggested that the reason why brood parasites such as cuckoos and cowbirds beg so loudly is that they are unconcerned with the costs experienced by their nest-mates. Studies of brood parasites are to some extent confounded by interspecific differences. However, one might predict that species of birds where intraspecific brood parasitism is common should have higher levels of begging. More speculatively, a particular nestling, hatched from an egg laid in a stranger's nest, might beg more loudly if it was able to recognize its condition.

(*b*) *Effects on evolution of clutch size*

We began this section by noting that in broods of identical offspring, the Lack clutch size is defined as the clutch size that maximizes the product of clutch size and average offspring fitness. However, we argued in the last section that sibling conflict may reduce offspring fitness. Sibling conflict will thus act as a selection pressure tending to lead to a reduction in clutch size. This occurs because a reduction in clutch size results in greater resource share per head and increased solicitation is less rewarding for a well-fed chick (recall the fitness of a chick rises with increasing resources but at a decelerating rate).

Figure 4 illustrates the interaction between sibling conflict and clutch size. We plot parental fitness as a function of brood size in three circumstances. To obtain curve (a), we assume that the young do not complete at all and that the parent thus 'wins' the

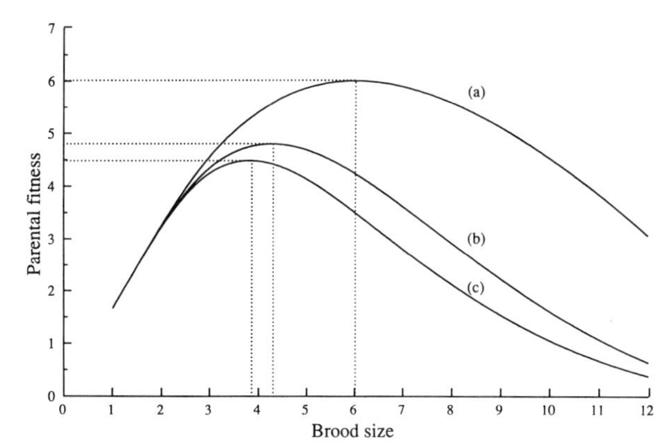

Figure 4. The relation between parental fitness and clutch size. (a) No sibling conflict, the optimum brood size is in this case 6. (b & c) The same relation but assuming that the young compete among themselves for resources and facultatively adjust their level of competition to clutch size; (b) summed or shared costs, (c) individual costs. (After Godfray & Parker 1991.)

conflict. To obtain the other two relations, we assume that the young compete among themselves at a rate appropriate to the size of the clutch in which they find themselves. Note, we are assuming a facultative response by offspring and parents to each other's strategies. Curves (b) and (c) differ as to how the costs of increased competition are spread among the brood members.

Several conclusions can be drawn from figure 4 (a full description and justification of the model is given by Godfray & Parker (1991). Sibling conflict reduces maximum parental fitness (the maxima of (b) and (c) are less than that of (a)) and leads to a reduction in the 'Lack clutch size'. The magnitude of the reduction depends on the nature of the distribution of costs: our results show that greater reductions occur with individual costs in comparison with summed or shared costs. Finally we note that the reduction in clutch, at least in this model, is quite large, of the order of 25–30 %.

In figure 3, we illustrated the relation between sibling conflict and brood size although assuming a fixed clutch size. The same relation can be calculated assuming that the observed clutch is the optimum *after* the parent has modulated its clutch size. It transpires (Godfray & Parker 1991) that the qualitative relation described in figure 3 is unaltered although the absolute level of sibling conflict is reduced.

We suggest that parent–offspring conflict reduces clutch size. Perhaps the best way to test this idea is to look for a relation between clutch size and within-clutch relatedness as the latter will influence the expected level of conflict. Higher levels of sibling conflict should result in selection pressures for greater reductions in clutch size. Thus one would predict that a within-species comparison of populations differing in the degree of intraspecific brood parasitism should reveal an inverse relation between the frequency of brood parasitism and clutch size.

Another test of the predictions is offered by some haplodiploid species that lay clutches of eggs that are either all male or all female. Parasitoid wasps in the

Table 1. *Brood size in the parasitoid genus* Achrysocharoides. *Data from Bryan (1981)*

	male broods		female broods	
	no.	size	no.	size
A. latreilli	112	1.0	111	2.0
A. zwoelferi	103	1.0	152	2.0
A. niveipes	432	1.0	344	2.2
A. cilla	406	1.2	445	1.9

genus *Achrysocharoides* (Chalcidoidea, Eulophidae) lay small clutches of eggs in the larvae of leaf-mining moths (Askew & Ruse 1974; Bryan 1983). For reasons that are not understood, four species in the genus lay either all-male or all-female broods (other species are either thelytokous (1), lay mixed broods (1) or are imperfectly studied (a number)). The average relatedness within all-male broods is $\frac{1}{2}$ while the average relatedness within all-female broods is $\frac{3}{4}$. After oviposition, the mother abandons her young and thus any sibling conflict will evolve to the offspring optimum. We thus predict greater conflict, and hence smaller clutch sizes in all male broods. The data for all four species support this prediction (table 1). Other parasitoid wasps also lay broods of one sex although by laying a single egg that divides asexually to produce a clutch of genetically identical individuals (polyembryonic reproduction). In these species, there should be no sibling conflict over resource share. It may be no coincidence that the largest clutches of any parasitoid wasp, by a factor of two, are found in polyembryonic species (Clausen 1940).

(c) *Sibling conflict and offspring hierarchies*

Our discussion of sibling conflict, at least in birds, has been flawed by the assumption that all offspring have identical competitive abilities and experience the same costs of competition. In fact, birds normally begin to incubate their eggs before completing their clutches and this leads to a hierarchy in chick sizes. It is highly likely that different size chicks have different competitive abilities and suffer different costs to begging. We first ask how these asymmetries may affect sibling competition and then go on to enquire whether the hierarchy itself may have arisen as an adaptation to reduce sibling conflict.

The most straightforward situation to analyse is the case where the largest chick has complete freedom to take as much resource as it wants; the second chick then takes as much as it wants of the remainder and so on down the size hierarchy. Parker *et al.* (1989) analysed a model incorporating these assumptions. Unlike the competitive resource division discussed above, this method of 'pre-emptive' resource division results in an evolutionarily stable distribution of resources between offspring without the need to invoke costs to the mechanism of division. In the two-chick case, the division of resource is determined solely by the behaviour of the elder chick. Its decision is based on the trade-off between using resources to increase its own

Table 2. *Division of resources between great egret chicks. A is the largest and C is the smallest chick. Data from Parker* et al. (*1989*)

	resource share	
	mean	s.d.
chick A	43.2%	6.5%
chick B	37.0%	3.6%
chick C	19.8%	5.8%

fitness against using the same resources to increase the fitness of its sibling. As the elder chick weights the fitness of its sibling by the coefficient of relatedness, the equilibrium resource share is biased in its favour. The process of division is fundamentally the same in larger broods except that now the calculations are more complicated as the optimal decision for the eldest chick will depend on the decisions made by smaller chicks: a set of simultaneous equations have to be solved to obtain the ESS.

One prediction of the model, that resource share should be related to the position in the hierarchy, is supported by data. Table 2 shows data collected by D. W. Mock and his colleagues on food share among chicks of the great egret (*Casmeroidius albus*). Not only does resource share decrease for smaller chicks, but the greatest discrepancy is between the second and third chicks, a feature also predicted by the model.

In reality, older chicks probably have an advantage, but not a hegemony, over resource division. We have explored a number of two-chick models of competitive resource division incorporating either (i) asymmetries in competitive abilities (see also Parker *et al.* 1989) or (ii) asymmetries in the costs of solicitation. Because of lack of space, we only summarize the results here and will present the full models elsewhere. We assume that the resource share obtained by an individual is a function of the difference between a measure of its begging efficiency and the average begging in the nest (i.e. mean matching). To study asymmetries in competitive ability, we weight the begging intensity of each chick by a factor related to its position in the hierarchy. Increased begging is more rewarding for large chicks than for small chicks. To study asymmetries in the costs of solicitation, we assume that the costs of begging are multiplied by a factor related to the position of the chick in the hierarchy. Smaller chicks suffer more from an increase in solicitation than do larger chicks.

Figure 5 illustrates some typical result from a model incorporating asymmetric competitive abilities†. As the asymmetries increase, the larger chick begs less. However, because of its competitive advantage, its fitness actually increases. In contrast, the rate of begging by the smaller chick increases as the asym-

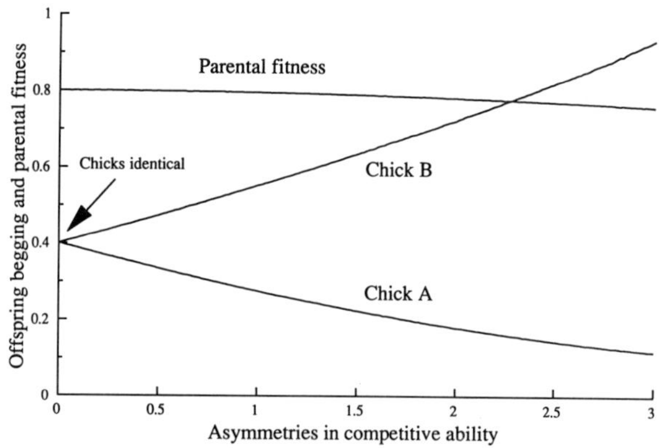

Figure 5. ESS levels of sibling begging with competitive asymmetry. Chick A has a competitive advantage over Chick B. In the absence of any asymmetry, the predicted level of begging (0.4) is identical to the two chick case of figure 3. As the asymmetry increases, the superior chick begs less and the inferior chick begs more. Parental fitness declines slowly from a figure about 0.8 times the value that could be achieved in the absence of sibling conflict.

metries widen while its fitness declines. Even when begging by the small chick is very inefficient, quite high levels of begging are still selected. To obtain figure 6, we assumed that there was no difference in competitive ability between the chicks but that the smaller chick suffered greater fitness penalties than the larger chick: apart from this the models are identical. The greater costs experienced by the small chick select for lower begging levels. The reduction in competition from its sibling allows the larger chicks also to reduce its level of begging.

The presence of sibling hierarchies can thus have important effects on resource distribution and on competitive solicitation. The evolution of sibling hierarchies has become a major preoccupation of experimental ornithologists and no less than eight separate hypotheses have been advanced as explanations (Lessells & Avery 1989; Slagsvold & Lifjeld 1989). One suggestion, its origins discernible in Hamilton (1964) but developed by Hahn (1981) and Mock & Ploger (1987), is that the hierarchy is an adaptation by the parent to reduce sibling competition.

We can explore this explanation using the models discussed in this paper. Consider the reduction in parental fitness which is caused by competitive begging in broods of identical siblings. In the particular model we used to obtain figure 3, the parent of a two-chick brood suffered a 20% drop in fitness owing to sibling competition. Suppose that the imposition by the parent of a hierarchy in chick sizes results in complete dominance of larger over smaller chicks, the pre-emptive form of resource division discussed above. The abnegation of resource division by the parent has a cost that depends on the rate at which offspring fitness increases when the offspring obtains extra resources above the parental optimum. By using a model exactly equivalent to that in figure 3, one can calculate that the reduction in parental fitness caused by pre-emptive resource division by the young is somewhere between 0% and 14% with values in the range 4–12% being perhaps the most realistic biologically. These figures,

† This model is a modification of the symmetric model in Godfray & Parker (1991) which was used to obtain figure 3. We have assumed individual costs and that the benefits of increased resources and the costs of solicitation combine multiplicatively. An additive model with individual costs (Parker *et al.* 1979) gives qualitatively very similar results as do multiplicative models with summed and shared costs (H. C. J. Godfray & G. A. Parker, unpublished results).

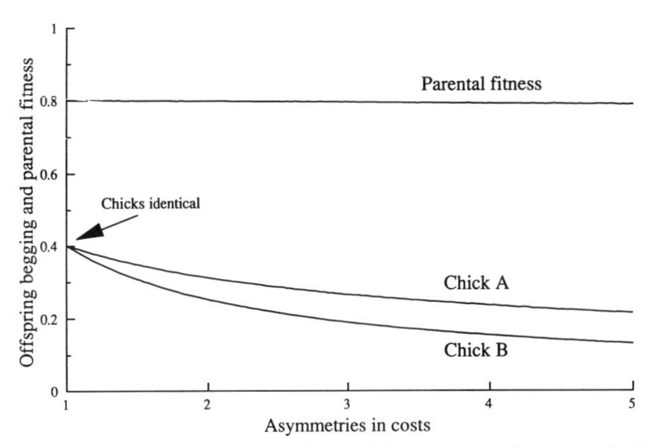

Figure 6. ESS levels of begging with asymmetric costs. As in figure 5 but increasing the asymmetry decreases begging by both chicks. Parental fitness again declines slightly with increasing asymmetry.

which should be compared with the 20% reduction caused by competitive solicitation, suggest an advantage to the parent of a hierarchy of chick sizes.

However, this conclusion is strongly contingent on the absence of any competitive solicitation. Consider the asymmetric solicitation models illustrated in figures 5 and 6. These models are again directly comparable with the symmetric model of figure 3. In both cases, parental fitness *declines* as the asymmetries between the young increase. We have explored several related models and have found a decline in parental fitness to the normal outcome though in some cases there may be a very small rise in parental fitness with increasing offspring asymmetry (Parker *et al.* 1989).

In conclusion, asymmetries among offspring affect the outcome of sibling competition. The parent may be selected to produce a hierarchy of chick sizes if this leads to pre-emptive resource division. However, we found little evidence that the introduction of asymmetries into models of resource division by competitive solicitation decreased the scope for parent–offspring conflict. To evaluate the importance of sibling competition in the evolution of chick hierarchies, further work is needed to discover exactly how resources are divided in nests of different sized chicks. In addition, the comparison of offspring and parent optima needs to be extended to inquire how the presence of a hierarchy affects the resolution of the conflict.

(*d*) *Siblicide*

We conclude this section by a discussion of the most direct way an offspring can affect clutch size: by killing one of its siblings. It has been known for some time that siblicide occurs in a number of bird groups and this phenomenon was referred to as 'Cain & Abel Conflict' by Gordon (1927). Siblicide is perhaps most frequent in large raptors although it also occurs in pelicans, gannets, cranes, skuas, penguins and herons (Lack 1968; O'Connor 1978; Drummond *et al.* 1986; Mock 1987; Anderson 1990). 83% and 76% of clutches of the lesser spotted eagle (*Aquila pomarina*) and the black eagle (*A. verreauxii*) contain two eggs yet in no case has a lesser spotted eagle been observed to rear two young (Mayburg 1973) and there is only a single possible case of twin fledging in the black eagle (Gargett 1970). In both cases siblicide is responsible for the reduction in clutch size. In a more typical eagle, the golden eagle (*A. chrysaetos*), sibling aggression always occurs although brood reduction only occurs in 80% of cases (Brown 1976). Siblicide also occurs in a number of invertebrate groups (Polis 1981) where it is frequently, although by no means always, associated with cannibalism.

Lack (1966) suggested that siblicide was an adaptation to allow the parent to respond to varying environmental conditions. In years of good food supply all chicks are reared whereas in poor years both the parent and the young benefit from a reduction in brood size through siblicide. In species with obligate, or near obligate siblicide, a parent may lay a second egg as an insurance against the infertility of the first egg (Dorward 1962; Anderson 1990).

O'Connor (1978) first pointed out that parents and young may differ over the conditions for brood reduction and that the extent of disagreement would depend on clutch size. There is, however, a potential problem in O'Connor's analysis as he used simple inclusive fitness arguments. Inclusive fitness arguments normally require weak selection with the trait having an additive effect on fitness, assumptions that may be violated by siblicide. These considerations suggest that siblicide may have to be analysed using explicitly genetic models. However, the assumption of additivity will remain valid in analyses of siblicide in large broods of animals where elder offspring kill a small number of their younger sibling. This behaviour, which occurs in a number of invertebrate groups such as chrysomelid and coccinelid beetles (Alexander 1974; Banks 1956), was modelled by Parker & Mock (1987) who indeed found disagreement between parent and young over clutch size. They estimated the magnitude of the disagreement by assuming biologically plausible relations between clutch size and fitness and concluded that, in general, the disagreement was likely to be small. In invertebrates with this type of siblicidal behaviour, the parent abandons its eggs after oviposition: it thus seems likely that in the absence of intervention by the parent, the offspring will 'win' the parent offspring conflict.

Despite the potential problems with O'Connor's analysis, explicit genetic models of sibicide in birds with small clutches confirms his results, although with one important proviso. The victim of siblicide must be a runt, unable to damage its siblings, even if it carries a siblicide gene (Godfray & Harper 1990)†. This is not

† A single locus analysis of siblicide produces an interesting theoretical curiosity. Over a wide range of parameter space both siblicidal behaviour and pure non-siblicidal behaviour are ESSs. However, only siblicidal behaviour is a continuous stable strategy (CSS, Eshel & Motro 1981; Eshel 1983). By definition, a resident strategy is an ESS if it can resist invasion by all mutant strategies (Maynard Smith 1892). However, suppose the resident strategy varies slightly, perhaps by chance: if this small deviation allows the spread of a mutant then the ESS is not a CSS. In the real world, we should only expect to observe those ESSs that are also CSSs. O'Connor identified a single ESS in his inclusive fitness analysis and this corresponds exactly to the CSS of the genetic analysis.

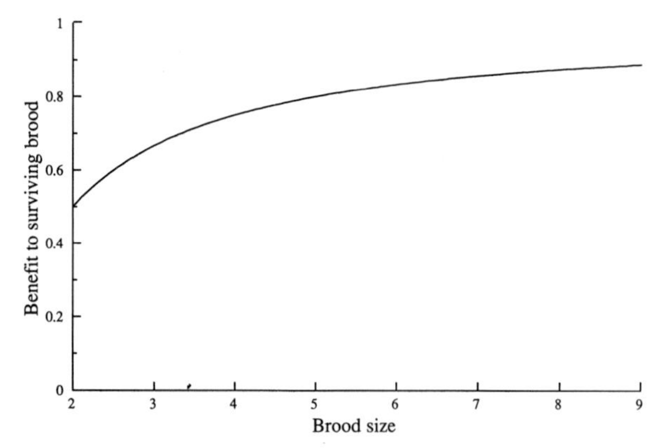

Figure 7. O'Connor's condition for brood reduction. Brood reduction is favoured by the young if the benefit to the surviving brood members is above the line. Benefit is measured in 'offspring equivalents'. Thus siblicide is favoured in a clutch of two if the fitness of the surviving chick increases by a factor of 0.5. The parent will favour siblicide if the benefit to the surviving brood members is greater than one.

an unreasonable assumption in most bird species where there is a hierarchy in chick sizes and an obvious runt. As O'Connor noted, the extent of parent–offspring conflict over the conditions for siblicide is greatest for small clutch sizes (figure 7).

In those cases of siblicide where there is not an identifiable and helpless victim, genetic models suggest that the presence of siblicidal behaviour will often depend on the past history of the population (Godfray 1987*a*; Godfray & Harper 1990). Models of this type probably apply to parasitoid wasps. The young of many parasitoid species have large mandibles which they use to kill all other larvae in the same host. Godfray (1987*a*) argued that the possession of fighting mandibles has some of the properties of an absorbing state, once evolved they are difficult to loose as any mutant, non-siblicidal allele is at a great disadvantage. This might explain why there is a dichotomy rather than a continuum in parasitoid clutch sizes: species tend either to lay single eggs (giving rise to fighting larvae) or clutches of eggs (non-fighting larvae) (Le Masurier 1987). It also suggests that there may be extensive parent–offspring conflict over clutch size. One wasp species, with fighting larvae, may be able to rear one individual from a particular host type, whereas a second a wasp species, of identical size but without fighting larvae, may be able to rear up to about 15 offspring from the same host.

5. PATTERNS IN LIFETIME REPRODUCTIVE SUCCESS

We now more from considerations of parental investment and clutch size to the wider question of the allocation of resources between reproductive and trophic functions over the lifetime of an organism. The trade-off between present and future reproductive success is one of the most fundamental problems of life-history theory and has been the subject of extensive analysis (Williams 1966; Stearns 1976; Sibly & Calow 1983). In this section we elaborate on a short note by Charnov (1982) who pointed out that parents and offspring will disagree about this division of resources, and that this may affect lifetime fertility and mortality schedules. Again our strategy is to locate offspring and parental optima to discover the scope for parent–offspring conflict. We present here a summary of some recent modelling that will be presented and justified in full elsewhere.

Consider an animal that produces one offspring a year throughout its life. The offspring benefits from increased investment by the parent, although with diminishing returns. However, increased investment by the parent in this season's offspring increases the probability of dying over the winter before the next breeding season. To simplify matters, we shall assume a strictly regulated population (by density-dependent recruitment to the breeding population) and that the only consequences of increased investment for the parent is decreased survival the following winter. Under these assumptions, the optimum parental investment can be found by maximizing the sum of present reproductive success and future reproductive success at each breeding season. (Technically, this is done by a process of backward iteration from a time sufficiently far in the future that the assumption can be made that future reproductive success is zero. By the time the process of iteration reaches an age to which the animal might be expected to have survived, the transient effects of any assumptions about the time horizon have disappeared.) The optimal allocation of resources for the offspring differs from that of its parent in that it devalues the future reproductive success of the parent by its coefficient of relatedness to future siblings.

In the simplest case, parental survival at any particular level of investment is constant and it is possible analytically to obtain the parent and offspring optima which are age independent. The amount of parent–offspring conflict depends, as before, on how much an offspring benefits from extra resources beyond the parental optimum. Figure 8 illustrates the drop in parental longevity and expected lifetime reproductive success that may occur if the offspring obtain complete control over resource allocation. The reduction in parental longevity and fitness is greatest when the offspring have most to gain from exceeding the optimum investment for the parent.

In reality, the probability of parental survival will vary with age as will the deleterious effects of increased investment. We have modelled such a situation by assuming an underlying sigmoid survival curve (we assumed a Weibull mortability distribution). In the statistical literature on survival analysis, the effects of extraneous variables on survival are frequently studied using an 'accelerated life model' (see for example, Cox & Oates 1984). High values of the variable increase mortality to a level that would otherwise have been experienced later in life, hence the term accelerated life. We have assumed that increased investment affects survival in this manner. Thus, if a five-year-old animal increases investment by one unit, it experiences a risk of mortality appropriate to (say) a six-year-old animal and, similarly, an eleven-year old animal would experience the risk of a twelve-year-old animal.

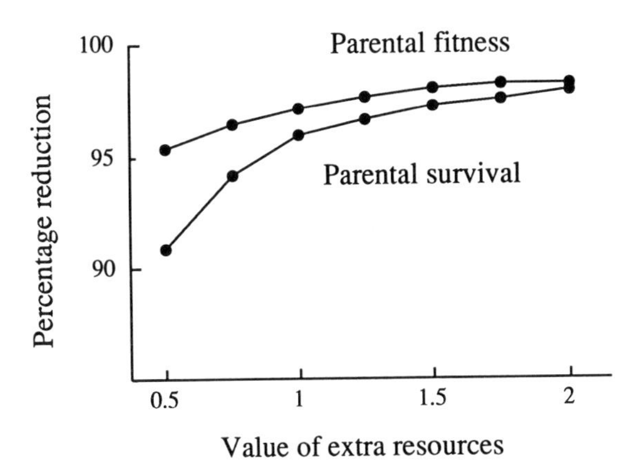

Figure 8. Parental fitness and fecundity (expressed as a percentage of the maximum) when the offspring determine the trade-off between present and future reproductive success. In the absence of changes in parental investment, mortality is constant throughout life. As in figure 2, these results were obtained using a specific relation between offspring fitness and resources described in Parker & Macnair (1978). The rate at which offspring fitness increases with additional resources beyond the parental optimum is determined by a single parameter which is plotted on the x axis. Low values imply a rapid gain of fitness and hence greater benefits for conflict.

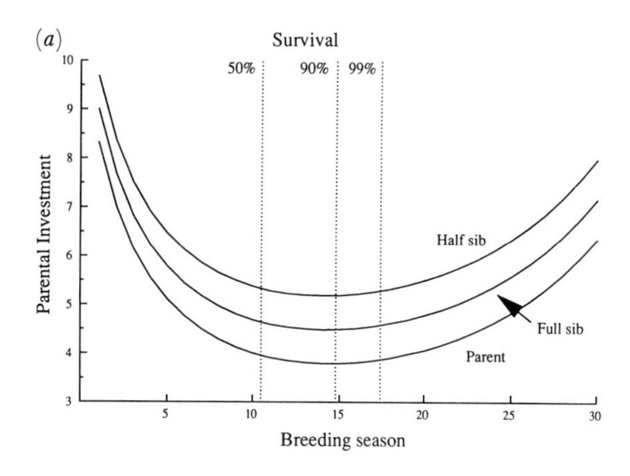

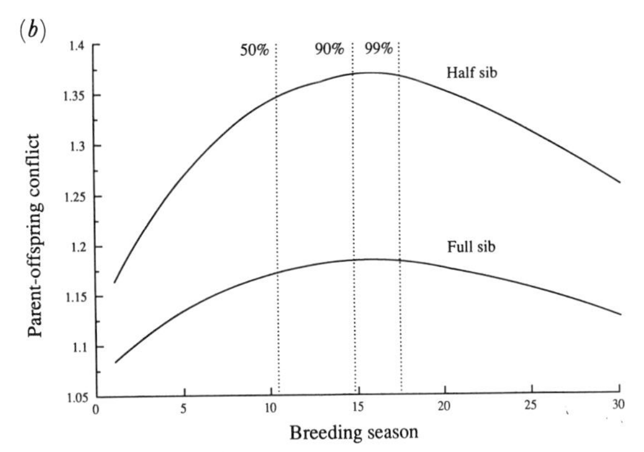

Figure 9. (*a*) Parental investment as a function of age when the background mortality is described by a Weibull distribution and extra parental investment acts to 'accelerate life'. The three curves describe investment when the decision is determined by the parent, a sibling or a half sibling. The times by which 50 %, 90 % and 99 % of an initial cohort will have died are shown by dotted lines. (*b*) The magnitude of parent–offspring conflict (offspring optimum minus parent optimum) as a function of parental age.

A typical result of this analysis is shown in figure 9. Annual reproductive investment is initially high but then drops as the animal ages. Only when the animal is very old does it begin to rise again. The extent of parent–offspring conflict appears to be inversely related to the quantity of parental investment. The amount of conflict rises as an animal ages but eventually falls in extreme old age. Note, it is very unlikely that either the rise in parental investment in old age, or the decrease in parent–offspring conflict would ever be observed: very few animals live long enough to manifest this trait.

To understand these results, recall that the effect of increased investment is to 'accelerate life', that is to increase the risk of mortality to that experienced by an older animal. As the risk of mortality when young is low, and remains low for several years, the consequences of increased investment are relatively mild. As the animal grows older and approaches the age when the risks of mortality accelerate, the consequences of increased investment become dire and this leads to a decrease in investment per offspring. When the animal is very old, the chances of surviving until the next breeding season are so small that the optimal investment in the current breeding season again begins to rise. This latter rise is a graded form of terminal investment.

Parent–offspring conflict increases with parental age for two reasons. First, as investment drops, the advantages of inveigling greater investment from the parent increases (recall the relation between offspring fitness and investment is monotonically increasing but with decelerating slope). In addition, the offspring is less concerned than its parents about the reduction in numbers of future siblings and is thus willing to sanction a level of investment that entails a greater risk of mortality for the parent.

6. CONCLUSIONS

Our aim in this paper has been to highlight the potential importance of parent–offspring conflict by comparing offspring and parent optima in a number of the classic problems from life-history theory. We have shown that such effects exist, and may indeed be substantial.

Our efforts will be of no more than technical interest if Alexander (1974) is right in his belief that the parent always succeeds in imposing its will on the offspring. However, this must be untrue in some cases, for example when a parent abandons a clutch of eggs after oviposition and is thus in no position to influence interactions between her young. Even when the parent remains with the clutch, it seems likely that parental attention to variation in the needs of the young will allow the young to demand and obtain more resources than is optimal for the parent. An obvious future step in the analysis of such problems is to follow Harper (1986) in developing sibling conflict models where offspring differ in their resource requirements. Recent developments in the study of animal signalling are also relevant here (Grafen 1990).

We also suggest that selection acting on the parent to avoid parent–offspring conflict may be an important factor in shaping reproductive strategies. We have quantitatively investigated the advantages to the parent of producing hierarchies of chick sizes and have found partial support for the suggestion that it may have evolved to reduce sibling conflict. We also suggest that sibling conflict may act as a selection pressure affecting clutch size. Without wanting to overestimate the importance of parent–offspring conflict, we do suggest that the avoidance of conflict should be considered when attempting to explain any facet of reproductive strategies, from clutch size to honey combs.

Finally, we stress that we are under no illusions that we have been dealing with real birds or insects. The models we have examined are general and tools for thought rather than suitable for immediate application to particular systems. Nevertheless, we do claim that our models generate qualitative predictions that are worth testing against nature. There are real problems (both ethical and practical) in parameterizing parent–offspring conflict models and performing manipulative tests: we believe that ingenuity in overcoming such problems could be richly rewarded. In addition, we believe that the massive literature on the reproductive biology of many animals, in particular birds, offers great prospects for comparative tests of predictions about parent–offspring conflict.

REFERENCES

Alexander, R. D. 1974 The evolution of social behaviour. *A. Rev. Ecol. Syst.* **5**, 325–383.

Alexander, R. D. & Sherman, P. W. 1977 Local mate competition and parental investment in social insects. *Science, Wash.* **196**, 494–500.

Anderson, D. J. 1990 Evolution of obligate siblicide in boobies. 1. A test of the insurance-egg hypothesis. *Am. Nat.* **135**, 334–350.

Askew, R. R. & Ruse, J. M. 1974 Biology and taxonomy of species of the genus *Enaysma* (Delucchi) (Hymenoptera: Eulophidae: Entedontinae) with special reference to the British fauna. *Trans. R. ent. Soc. Lond.* **125**, 257–294.

Banks, C. J. 1956 Observations on the behaviour and mortality in Coccinellidae before dispersal from the eggshells. *Proc. R. ent. Soc. Lond. A* **31**, 56–60.

Bengtsson, H. & Ryden, O. O. 1981 Development of parent-young interaction in asynchronously hatched brood of altricial birds. *Z. Tierpsychol.* **56**, 255–272.

Boomsma, J. J. 1989 Sex-investment ratios in ants: has female bias been systematically overestimated. *Am. Nat.* **133**, 517–532.

Boomsma, J. J. & Grafen, A. 1990 Intraspecific variation in ant sex ratios and the Trivers–Hare hypothesis. *Evolution* **44**, 1026–1034.

Brown, L. H. 1976 *British birds of prey*. London: Collins.

Bryan, G. 1983 Seasonal biological variation in some leaf-miner parasites in the genus *Achrysocharoides* (Hymenoptera, Eulophidae). *Ecol. Ent.* **8**, 259–270.

Charnov, E. L. 1982 Parent–offspring conflict over reproductive effort. *Am. Nat.* **119**, 736–737.

Charnov, E. L. & Krebs, J. R. 1974 On clutch size and fitness. *Ibis* **116**, 217–219.

Clausen, C. P. 1940 *Entomophagous insects*. New York: McGraw-Hill.

Clutton-Brock, T. H. 1991 *Parental investment*. Princeton University Press.

Clutton-Brock, T. H. & Godfray, H. C. J. 1991 Parental Investment. In *Behavioural ecology*, 3rd edn (ed. J. R. Krebs & N. B. Davies). Oxford: Blackwell Scientific Publications. (In the press.)

Cox, D. R. & Oates, D. 1984 *Analysis of survival data*. London: Chapman & Hall.

Dawkins, R. 1976 *The selfish gene*. Oxford University Press.

Dorward, D. F. 1962. Comparative biology of the white booby and brown booby *Sula* spp., at Ascension. *Ibis* **103b**, 174–200.

Drummond, H., Gonzalez, E. & Osorno, J. L. 1986 Parent-offspring cooperation in the blue-footed booby (*Sula nebouxii*): social roles in infanticidal brood reduction. *Behav. Ecol. Sociobiol.* **19**, 365–372.

Eshel, I. 1983 Evolutionary and continuous stability. *J. Theor. Biol.* **103**, 99–112.

Eshel, I. & Motro, U. 1981 Kin selection and strong evolutionary stability of mutual help. *Theor. Pop. Biol.* **19**, 420–433.

Feldman, M. W. & Eshel, I. 1982 On the theory of parent-offspring conflict: a two-locus genetic model. *Am. Nat.* **119**, 285–292.

Gargett, V. 1978 Sibling aggression in the Black Eagle in the Matapos, Rhodesia. *Ostrich* **49**, 57–63.

Godfray, H. C. J. 1987*a* The evolution of clutch size in parasitic wasps. *Am. Nat.* **129**, 221–233.

Godfray, H. C. J. 1987*b* The evolution of invertebrate clutch size. *Oxford Surv. Evol. Biol.* **4**, 117–154.

Godfray, H. C. J. & Harper, A. B. 1990 The evolution of brood reduction by siblicide in birds. *J. theor. Biol.* **145**, 163–175.

Godfray, H. C. J. & Parker, G. A. 1991 Sibling competition, parent–offspring conflict and clutch size. *Anim. Behav.* (In the press.)

Gordon, S. 1927 *Days with the golden eagle*. London: William & Norgate.

Grafen, A. 1990 Biological signals as handicaps. *J. theor. Biol.* **144**, 517–546

Hahn, D. C. 1981 Asynchronous hatching in the laughing gull: cutting losses and reducing sibling rivalry. *Anim. Behav.* **29**, 421–427.

Haig, D. 1987 Kin conflict in seed plants. *Trends Ecol. Evol.* **2**, 337–340.

Hamilton, W. D. 1964 The genetical theory of social behaviour, I & II. *J. theor. Biol.* **7**, 1–16 & 17–51.

Hamilton, W. D. 1967 Extraordinary sex ratios. *Science, Wash.* **156**, 477–488.

Harper, A. B. 1986 The evolution of begging: sibling competition and parent-offspring conflict. *Amer. Nat.* **128**, 99–114.

Hausfater, G., Aref, S. & Cairns, S. J. 1982 Infanticide as an alternative male reproductive strategy in langurs: a mathematical model. *J. theor. Biol.* **94**, 391–412.

Lack, D. 1947 The significance of clutch size. *Ibis* **89**, 309–352.

Lack, D. 1966 *Population studies of birds*. Oxford: Clarendon Press.

Lack, D. 1968 *Ecological adaptations for breeding in birds* London: Methuen.

Lazarus, J. & Inglis, B. 1986 Shared and unshared parental investment, parent-offspring conflict, and brood size. *Anim. Behav.* **34**, 1791–1804.

Lessells, C. M. & Avery, M. I. 1989 Hatching asynchrony in European bee-eaters *Merops apiaster*. *J. Anim. Ecol.* **58**, 815–836.

Lloyd, D. 1987 Selection of offspring size at independence and other size versus number strategies. *Am. Nat.* **129**, 800–817.

Macnair, M. R. & Parker, G. A. 1978 Models of parent-offspring conflict. II. Promiscuity. *Anim. Behav.* **26**, 111–122.

Macnair, M. R. & Parker, G. A. 1978 Models of parent-offspring conflict. III. Intra-brood conflict. *Anim. Behav.* **27**, 1202–1209.

Le Masurier, A. D. 1987 A comparative study of the relationship between host size and brood size in *Apanteles* spp. (Hymenoptera, Braconidae). *Ecol. Ent.* **12**, 383–393.

Maynard Smith, J. 1982 *Evolution and the theory of games.* Cambridge University Press.

Meyburg, B.-U. 1974 Sibling aggression and mortality among nestling eagles. *Ibis* **116**, 224–228.

Mock, D. W. 1987 Siblicide, parent-offspring conflict, and unequal parental investment by egrets and herons. *Behav. Ecol. Sociobiol.* **20**, 247–256.

Mock, D. W. & Ploger, B. J. 1987 Parental manipulation of optimal hatch asynchrony in cattle egrets: an experimental study. *Anim. Behav.* **35**, 150–160.

Nonacs, P. 1986 Ant reproductive strategies and sex allocation theory. *Q. Rev. Biol.* **61**, 1–21.

O'Connor, R. J. 1978 Brood reduction in birds: selection for fratricide, infanticide and suicide? *Anim. Behav.* **26**, 79–96.

Parker, G. A. 1985 Models of parent-offspring conflict. V. Effects of the behaviour of two parents. *Anim. Behav.* **33**, 519–533.

Parker, G. A. & Begon M. 1986 Optimal egg size and clutch size: effects of environment and maternal phenotype. *Am. Nat.* **128**, 573–592.

Parker, G. A. & Courtney, S. P. 1984 Models of clutch size in insect oviposition. *Theor. Popul. Biol.* **26**, 27–48.

Parker, G. A. & Macnair, M. R. 1978 Models of parent-offspring conflict. I. Monogamy. *Anim. Behav.* **26**, 97–110.

Parker, G. A. & Macnair, M. R. 1979 Models of parent-offspring conflict. IV. Suppression: Evolutionary retaliation by the parent. *Anim. Behav.* **27**, 1210–1235.

Parker, G. A. & Mock, D. W. 1987 Parent-offspring conflict over clutch size. *Evol. Ecol.* **1**, 161–174.

Parker, G. A., Mock, D. W. & Lamey, T. C. 1989 How selfish should stronger sibs be? *Am. Nat.* **133**, 846–868.

Polis, G. A. 1981 The evolution and dynamics of intraspecific predation. *A. Rev. Ecol. Syst.* **12**, 225–251.

Ricklefs, R. E. 1969 An analysis of nesting mortality in birds. *Smithson. Contr. Zool.* **9**, 1–112.

Ryden, O. O. & Bengtsson, H. 1980 Differential begging and locomotory behaviour by early and late hatched nestlings affecting the distribution of food in asynchronously hatched broods by altricial birds. *Z. Tierpsychol.* **53**, 209–224.

Skutch, A. F. 1949 Do tropical birds rear as many young as they can nourish: *Ibis* **91**, 430–455.

Sibly, R. M. & Calow, P. 1983 An integrated approach to life-cycle evolution using selective landscapes. *J. theor. Biol.* **102**, 527–547.

Stamps, J. A., Metcalf, R. A. & Krishman, V. V. 1978 A genetic analysis of parent-offspring conflict. *Behav. Ecol. Sociobiol.* **3**, 367–392.

Slagsvold, T. & Lifjeld, J. T. 1989 Constraints on hatching asynchrony and egg size in pied flycatchers. *J. Anim. Ecol.* **58**, 837–850.

Smith, C. C. & Fretwell, S. D. 1974 The optimal balance between size and numbers of offspring. *Am. Nat.* **108**, 499–506.

Stearns, S. C. 1976 Life-history tactics: a review of the ideas. *Q. Rev. Biol.* **51**, 3–47.

Trivers, R. L. 1974 Parent-offspring conflict. *Am. Zool.*, **14**, 249–264.

Trivers, R. L. & Hare, H. 1976 Haplodiploidy and the evolution of the social insects. *Science, Wash.* **191**, 249–263.

Williams, G. C. 1966 *Adaptation and natural selection.* Princeton University Press.

Winkler, D. W. & Wallin, K. 1987 Offspring size and number: a life history model linking effort per offspring and total effort. *Am. Nat.* **129** 708–720.

Discussion

D. Haig (*Department of Plant Sciences, Oxford University, U.K.*). In your paper, you give the relatedness among half sibs as a quarter. This is actually the average of a relatedness of a half for maternal genes in the offspring, and a relatedness of zero for the paternal genes. Whenever there are interactions among half sibs, there is potential for conflict between maternal and paternal genes within offspring, and a possibility that natural selection will result in alleles that display differential gene expression depending on the parent of origin (i.e. 'genomic imprinting'). For example, in eutherian mammals, the paternal genome appears to be particularly active in the development of extraembryonic membranes. If so, the averaging of maternal and paternal relatedness in models of parent–offspring conflict may give misleading quantitative predictions.

H. C. J. Godfray. Our assumption that the relevant measure of relatedness between half sibs is a quarter would be wrong if both (i) maternal and paternal genes were capable of independently influencing the behaviour of offspring and (ii) the outcome of conflict between the two types of genes was the complete, or predominant, control by either the maternal or paternal set. We think that independent action by maternal and paternal genes is most likely to be important at the embryonic stage where gene products still have very direct effects on growth and development. In addition, we think that the most likely outcome of conflict between the two sets of genes is a compromise where the assumption of a coefficient of relatedness of a quarter would still apply, at least approximately. However, this is an extremely interesting point and deserves further exploration.

The architecture of the life cycle in small organisms

GRAHAM BELL AND VASSILIKI KOUFOPANOU

Biology Department, McGill University, 1205 Avenue Dr Penfield, Montreal, Quebec, Canada H3A 1B1

SUMMARY

The life cycle of eukaryotes has a dual nature, composed of a vegetative cycle of growth and reproduction, and a sexual cycle of fusion and reduction, linked by the spore. Large size is often favoured through interactions with other organisms, or as a means of exploiting locally or temporarily abundant resources, despite the metabolic penalty of size increase. Beyond a certain point, large organisms must be multicellular (or multinucleate) because of the requirement for more deoxyribonucleic acid (DNA) to service larger quantities of cytoplasm. Multicellularity evolves in some lineages but not in others because its evolution is constrained by the pattern of spore development, being favoured, for example, by the occurrence of multiple fission as the consequence of possessing a rigid cell wall. The separation of soma from germ is also the outcome of a developmental constraint, in this case the inability of cells to divide while flagellated, and also the necessity of remaining in motion. Once achieved, a general physiological advantage is realized through the specialization of soma as source and germ as sink. Large, complex multicellular organisms are fragile constructs that can only persist through deploying sophisticated devices for maintenance. Thus two crucial, and related, properties of life cycles are repair and repeatability. The dual life cycle achieves exogenous repair through spore production in the vegetative cycle and through outcrossing and recombination in the sexual cycle. Repeatability is enhanced by developmental mechanisms such as maternal control and germ-line sequestration, which by restricting the occurrence or the heritability of somatic mutations promote their own replication.

1. INTRODUCTION

(*a*) *Dual nature of the life cycle*

All organisms pass through a more or less stereotyped sequence of changes before returning to some initial state. This process is called the life cycle, to emphasize its cyclical, or repetitive, nature. The concept is closely related to those of the life history, development and ontogeny, but differs from them in stressing renewal. It expresses the wonder with which people have observed that old organisms give birth to young organisms, and continue to do so, generation after generation, apparently without limit.

Figure 1 shows a general representation of the life cycle, omitting many of the features of individual cases, but showing the relation between the most important processes. Our interpretation is based on two concepts. The first is the central position of the simple spore cycle: a unicellular spore which grows in size before dividing into two similar spores. Many protists actually possess such a cycle, but it can also be used to imagine what modifications are involved in producing more elaborate life cycles. The second is the dual nature of the life cycle, which we have divided into a vegetative cycle and a sexual cycle. The vegetative cycle involves growth and reproduction, and is responsible for change in the quantity of substance. The sexual cycle involves fusion and reconstitution, and produces change in quality. The two cycles are linked by the spore, which can be switched into either vegetative or sexual expression, knitting together the two cycles into a single life cycle. No aspect of the biology of life cycles is more striking than the fact that vegetative proliferation is very seldom, if ever, the sole mode of reproduction, the lineage instead being reduced at more or less frequent intervals to a single cell, whether produced sexually or asexually. One explanation of this remarkable pattern is that large mass of tissue will inevitably bear, in different cells, a variety of somatic mutations, which, as they cannot be eliminated, must continue to accumulate generation after generation. By producing many independent germ cells an individual ensures that any which bear somatic mutations are likely to be eliminated by selection. Adaptations of this sort, in which parental fitness is enhanced by competition among many cheap offspring, have been called 'selection arenas' by Kozlowski & Stearns

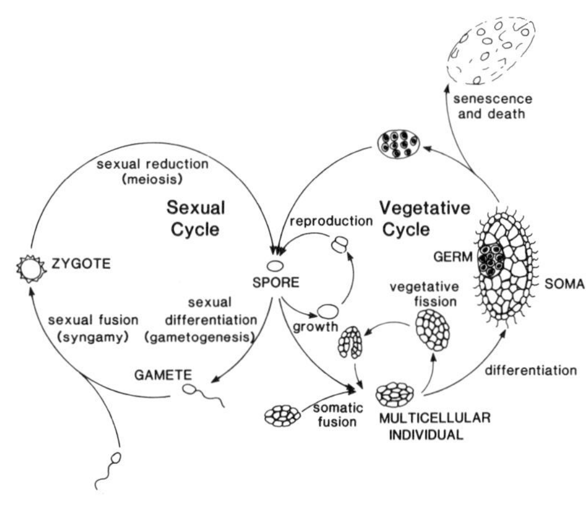

Figure 1. The life cycle.

Phil. Trans. R. Soc. Lond. B (1991) **332**, 81–89
Printed in Great Britain

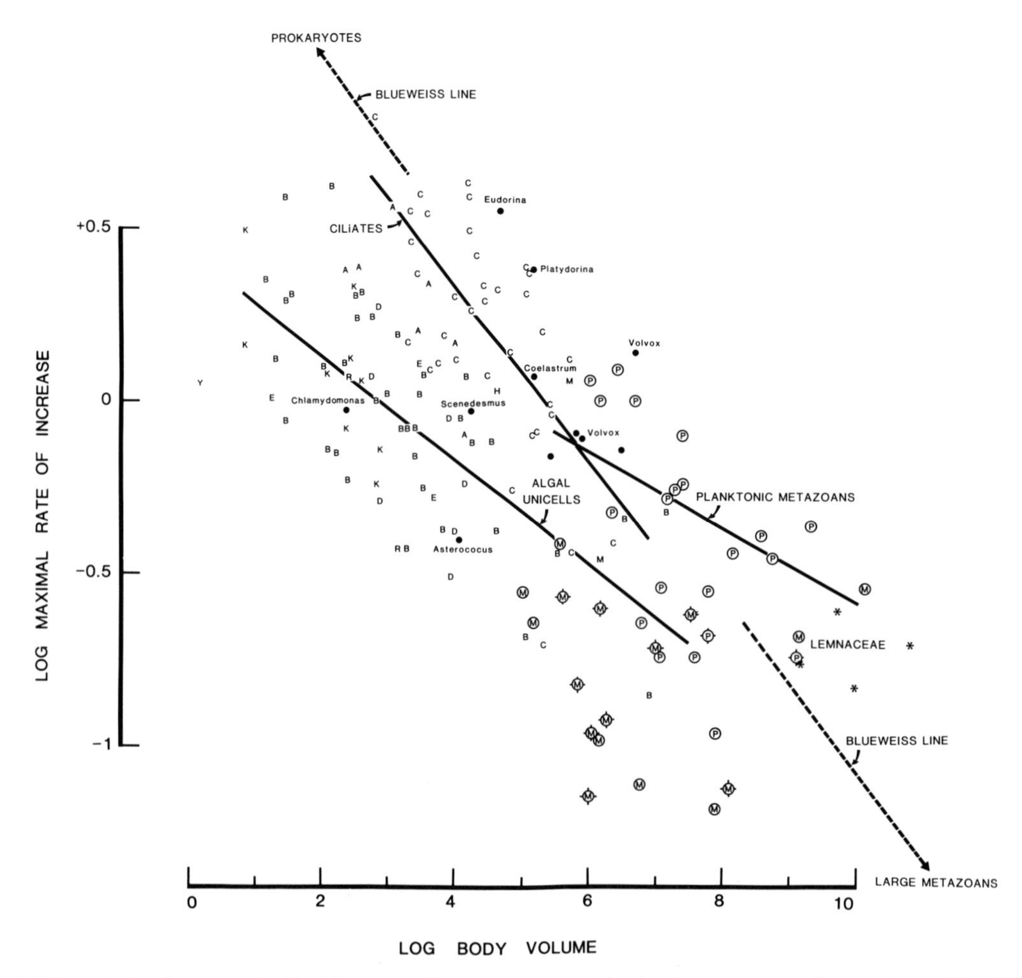

Figure 2. The relation between the limiting rate of increase r_{max} and body size among small organisms. The different taxa shown are: A, amoebas; B, Bacillariophyceae (diatoms); C, ciliates; D, Pyrrhophyceae; E, euglenids; H, Heliozoa; K, Chlorophyceae; M, benthic metazoans; P, planktonic metazoans; R, Cryptophyceae; Y, Myxophyceae. Metazoans are distinguished by a circle surrounding the letter; when the circle is pecked the organism is exclusively sexual, and r_{max} should be multiplied by two (add 0.3 on the log scale of the graph) to obtain values comparable with the remaining, asexual organisms. Solid lines are least-squares regressions for planktonic metazoans (P), ciliates (C) and algal unicells (B, D, E, K). The broken line is the regression obtained by Blueweiss *et al.* (1978), by using a smaller number of species over a wider size range. From Bell (1985).

(1989). A related idea is that somatic tissues become infected with pathogens, such as viruses, of which the lineage can be cleansed only through the production of numerous unicellular propagules, some of which will be uninfected. The notion that the life cycle facilitates exogenous repair by exposing offspring to natural selection will be developed in more detail below.

In this paper we have attempted to interpret the major features of life cycles shown in figure 1, by reference to small and simple organisms. The discussion is based on work done in our own laboratory on ciliates, *Chlamydomonas*, and multicellular green algae; it is not intended to be a review of protistan life cycles. Most of the theoretical issues that we address have been dealt with more extensively in recent books by Buss (1987), Bell (1989) and Bonner (1990).

2. THE VEGETATIVE CYCLE

(*a*) *Selection for small size*

The simplest life cycle comprises growth from initial size m to final size M, followed by complete fission into M/m propagules. The optimal values of m and M depend on the relation of fitness to size. If the environment is uniform and invariant, and if there are no interactions among individuals, then fitness can be equated with the rate of increase r in pure culture. This may take any value, depending on the physical conditions of culture; however, the maximum value of r which can be attained under nearly ideal conditions in the laboratory, r_{max}, is greatest for small organisms and declines with increasing size (Blueweiss *et al.* 1978). This generalization seems to hold at all scales; figure 2 shows the decline in rates of increase with size among small protists and metazoans. The optimal form of this cycle is thus to make m as small as possible, to grow to $M = 2m$, and then to divide by an equal binary fission. This is a common, but by no means a universal, pattern among prokaryotes and small unicellular eukaryotes.

The source of the disadvantage of increasing size seems to be related primarily to the smaller ratio of surface area to volume in larger cells, and thus to the lower specific rates of movement of materials and energy across surfaces. It is therefore rather surprising that eukaryotic unicells are not always as small as possible, and do not always reproduce by binary fission. This introduces three questions that are fundamental to understanding the elaboration of the asexual cycle. First, what are the benefits of greater size that would cause it to be favoured by selection? Secondly, how are large organisms constructed so as to minimize the ineluctable penalty of size increase? And thirdly, why does large size evolve in some lineages and not in others?

(*b*) *Evolution of large size*

The benefits of large size and multiple fission arise from the failure of the assumptions that lead us to infer that small size is generally advantageous.

The environment may vary in space. If some sites are better than others then it will pay to exploit a favourable site by prolonged growth before producing dispersive propagules. Algal mats in sunny shallows and parasites that have found hosts are very familiar examples of this tendency. We have a strong although unquantified impression that continued growth and multiple fission are much more common among benthic, attached and parasitic protists than among free-living planktonic forms. Ciliates provide a good example of a group in which freeswimming forms are invariably unicellular whereas attached forms often develop into large colonies. Environmental heterogeneity provides a simple explanation of such patterns.

The environment may vary in time. If resources are often scarce, larger organisms will take longer to starve to death. This is simply the inverse of the argument that smaller organisms have higher specific metabolic rates. It is the foundation of the 'size-efficiency' hypothesis (Brooks & Dodson 1965), which states that larger organisms of given form have lower threshold concentrations of food for growth and reproduction.

Interactions among organisms are ubiquitous, and larger organisms may generally prevail in antagonistic interactions. It is obvious enough that larger organisms can eat smaller ones, and are themselves less likely to be eaten.

Spatial heterogeneity, temporal variation and antagonistic interaction provide three general theories of size increase. Whether or not such processes create a net benefit for increased size in some particular case will further depend on the structure of the organism concerned. Pre-existing design thus represents a further constraint on the evolution of large size. An example is the evolution of motile multicellular forms among walled but not among naked green algae, which is discussed at greater length below.

(*c*) *Body form of large organisms*

Starting from a small, essentially point-like unicell, larger organisms can be constructed in one of three ways. Extension in one dimension produces a filament, as in fungi and filamentous algae. This design maximizes contact with the local environment and is thus specialized for absorption. Extension in two dimensions produces a sheet, as in macroalgae, bryophytes and tracheophytes. This design maximizes the interception of fluxes, typically of light. Not all plants are simple sheets, but their assimilatory organs, such as leaves, are usually sheet-like, or else their structure or behaviour are modified so as to create an essentially two-dimensional form, such as the spiral chloroplast of gamophytes, or the rotation of large motile algae. Extension in three dimensions produces a massive body, as in ciliates and animals. This design permits the enclosure of other organisms and is thus specialized for ingestion.

(*d*) *Internal structure and the nucleoplasmic ratio*

All of these body plans incur the penalty of decreased specific rates of metabolism, although this will be least for filaments and greatest for massive bodies. In particular, a cell that is maintaining and increasing a large bulk of cytoplasm requires high rates of protein synthesis, which in turn requires large quantities of RNA and DNA. Because greater quantities of DNA are replicated more slowly, there is a tension between the demands of growth and those of reproduction.

The design of large cells and organisms is thus strongly influenced by the need to replicate large quantities of DNA as rapidly as possible. The simplest solution is polyploidy, as in large amoebas. This seems to be associated with low rates of replication: polyploid amoebas divide more slowly than other unicells of comparable size, perhaps because the nuclear surface area increases less rapidly than nuclear mass. A second solution is to become multinucleate. This is very common among large protists such as amoebas, ciliates and foraminiferans, and also occurs in fungi, and even in metazoans such as orthonectids. Without further refinement, however, the multinucleate habit makes it difficult to synchronize nuclear and cell division, and, more fundamentally, introduces the possibility of reproductive competition among nuclei that may result in the spread of selfish nuclei to the detriment of organismal function (Buss 1987). A final solution is to become multicellular. This can be achieved either by the failure of cells to separate after division, as in most multicellular algae, plants or animals, or by the reaggregation of separate cells, as in cellular slime moulds, sponge gemmules, or the capsular embryos of some turbellarians. A body comprising many similar quasi-autonomous cells is successful for the same reason that cells containing many similar quasi-autonomous organelles such as mitochondria are successful: the design ensures that each volume of cytoplasm is effectively served by a nearby centre of metabolism.

(*e*) *The division of labour within large organisms*

Multinucleate and multicellular organisms often further reduce the penalties of large size by a cooperative division of labour among dissimilar structures.

Large unicells must reconcile conflicting requirements for a large genome to support vegetative metabolism and a small genome to allow rapid replication. Most ciliates and forams have achieved this through the separation of somatic and germinal functions between macronucleus and micronucleus. The ciliate macronucleus is typically a large, highly polyploid nucleus capable of sustaining vegetative metabolism and dividing rapidly but imprecisely by amitosis. In large ciliates the macronucleus is often elaborately shaped so as to increase the surface area in contact with the cytoplasm. The micronucleus is a smaller diploid nucleus which does not participate in vegetative metabolism but is replicated with great precision by a regular mitosis. This division of labour between somatic and germinal nuclei explains why ciliates are able to sustain high rates of increase despite their large size (see figure 2).

In multicellular organisms, labour is divided among cells, the fundamental distinction again lying between somatic and germinal functions. The relation between soma and germ is the subject of the next section.

3. THE VEGETATIVE CYCLE IN THE VOLVOCALES

(*a*) *The evolution of motile multicellular algae*

The quantitative relations between soma and germ are best known in the Volvocales, a group of flagellated green algae which spans the entire range from unicells to differentiated multicellular individuals. The material on which the following sections are based is described more fully by Bell (1985) and Koufopanou (1991).

The volume of motile green unicells ranges over three orders of magnitude, from about 10^1 to about 10^4 μm^3. Larger forms are colonial, and extend the size range up to about 10^6–10^7 μm^3 in algae comprising several thousand cells. For comparison, the largest immotile green unicells, the desmids, have volumes of up to 10^7 μm^3, whereas large attached forms such as *Ulva* have protoplast volumes of roughly 10^{13} μm^3; motile forms are thus relatively small. The multicellular habit has arisen at least six times in the Volvocales. The simplest colonies, found in Dunaliellaceae and Polyblepharidaceae, are loose aggregations of an indeterminate number of cells; such colonies reproduce by fragmentation, subsequently increasing in bulk through the binary fission of each of their members. Being associated with other cells in a colony probably offers some protection against filter-feeders such as rotifers. The more profound shift in design to regular multicellular individuals ('coenobia') is associated with the presence of a cell wall. Naked and scaly unicells invariably reproduce by binary fission and form loose colonies that reproduce by fragmentation. Unicells that have a rigid cell wall (such as *Chlamydomonas*) form 2–16 spores, which are liberated when the parental cell wall breaks down. Multiple fission presumably represents an economy of scale associated with the fixed overhead cost of discarding the cell wall. Colonies of such forms comprise a fixed number of cells arranged in a characteristic manner, and each cell in the colony divides so as to create the appropriate number of cells, constituting a miniature colony after which liberation enlarges by the equal growth of all its cells ('autocolony formation'). The evolution of regular multicellular individuals is thus contingent on the prior possession of a particular mode of development ('eleutheroschisis') involving repeated division within a rigid cell wall.

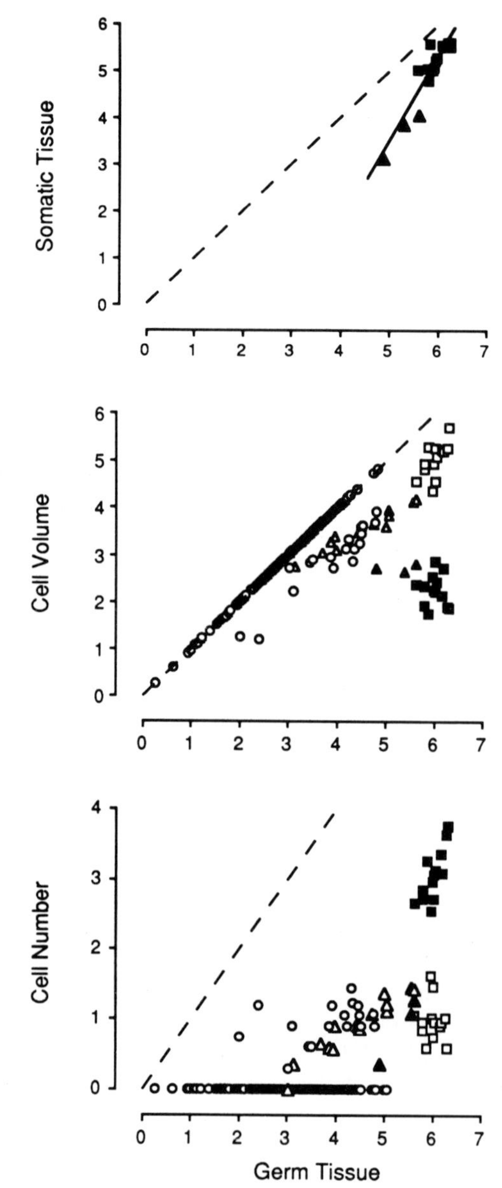

Figure 3. The quantitative relation between somatic and germinal tissue in flagellated green algae. From Koufopanou (1991). Solid symbols represent soma (triangles, *Pleodorina*; squares, *Volvox*) and open symbols germ tissue (circles, unicells and non-Volvocacean colonies; triangles, Volvocaceae except *Volvox*; squares, *Volvox*).

Colonies of this sort comprise 4–64 identical cells, arranged as a plate, ball or hollow sphere, and usually embedded in a common matrix. Among unicells, there is a steep regression of daughter cell volume on parental cell volume: larger forms divide to give larger daughters. Among coenobia there is no relation between the sizes of daughter and parental cells: larger colonies develop because larger parents undergo more divisions to form more daughters and thus a coenobium with more cells (figure 3). The evolution of larger multicellular individuals among the Volvocales thus proceeds by heterochronic shifts in the timing of the initiation of cell division.

(*b*) *Soma and the flagellation constraint*

In coenobia with 64–128 cells, in *Eudorina* and more distinctly in *Pleodorina*, there first appears a distinction between two cell lineages, a sterile soma and a fertile germ. The soma is first manifested as a tier of small anterior cells, and is thus associated with directional movement, in a manner reminiscent of metazoan amphiblastulae. The emergence of a separate caste of sterile cells appears to follow from a constraint on flagellation, as suggested by Buss (1987; see also Margulis & Sagan 1986). Naked green unicells retain the parental flagella during cell division, and so remain motile throughout their life cycle. Walled unicells, however, must discard their flagellum when dividing, being constrained mechanically by the presence of the cell wall and physiologically by possessing only a single centre of developmental activity, responsible both for cell division and for flagellation. Volvocacean colonies follow the same mode of development as their ancestral walled unicells: the basal bodies move away from the flagellar bases during cell division, and the flagella would consequently be unable to keep the large colonies from sinking during the long period of their development.

In larger forms such as *Volvox*, with many hundreds or thousands of cells, the distinction between soma and germ becomes steadily more exaggerated. The evolution of such forms does not proceed by heterochrony, but by shifts in the allocation to soma and germ caused by unequal cell divisions during development (see Kirk & Harper 1986).

(*c*) *Soma and germ as source and sink*

The separation of soma from germ realizes an important physiological advantage, which is shown by the fact that large coenobia have a greater maximal rate of increase than do unicells of comparable size (figure 2). Bell (1985) has suggested that the cause of this advantage is the specialization of the soma as a source and the germ as a sink of metabolites. The rate of chemical reactions, including photosynthesis, is generally reduced by the accumulation of end-products. In unicells, the intracellular concentration of products such as starch is reduced by their localization in structures such as the pyrenoid. Differentiated multicellular individuals can go further by translocating substances from the somatic cells in which they are produced to the germ cells where they will be used for reproduction and embryogenesis. In this way the rate of synthesis is increased by steepening the concentration gradient between the external medium and the cytoplasm of the soma. We have attempted to test this theory by the following experiment, which is described in more detail by Koufopanou & Bell (1991*a*).

The asexual germ cells, or gonidia, of *Volvox* will develop normally if isolated from the parental coenobium and cultured in an appropriate medium. We can make three predictions about the effect of the culture medium on the rate of growth of gonidia. First, if growth is limited by the supply of nutrients then the rate of growth, both of isolated gonidia and of those within intact parental coenobia, will increase with the concentration of the medium. Secondly, if the soma supplies resources to the developing gonidia, then those in intact coenobia will grow more rapidly than those which are cultured as isolated cells. Finally, if the source-and-sink hypothesis is correct, then the difference in growth rate between isolated gonidia and those in intact coenobia will be least at low nutrient concentrations, where feedback inhibition will be minimal, and greatest at high nutrient concentrations, where the advantage of a differentiated body should be expressed most clearly. The result of the experiment was consistent with all three predictions. In more natural circumstances, the eutrophication of ponds is associated with an increase in the frequency of colonial forms (Koufopanou & Bell 1991*a*).

The evolution of multicellularity in the Volvocales thus illustrates the interplay between general adaptive trends and specific developmental constraints. Multicellularity evolves readily because the possession of a rigid cell wall favours multiple fission, which permits the reproduction of regular multicellular individuals by autocolony formation. Large coenobia must remain motile while dividing, but because they possess only a single microtubule-organizing centre can do so only by setting aside a lineage of sterile flagellated somatic cells. The specialization of somatic cells as source and of germ cells as sink realizes an important physiological advantage in some ecological conditions, explaining why differentiated coenobia can reproduce faster than unicells of comparable size.

(*d*) *The allometry of soma and germ*

The quantitative relation between soma and germ is shown in figure 3. The total quantity of somatic tissue rises steeply with the quantity of germ. This may set an upper limit to the size of organisms with this body-plan, and the figure seems to show why forms larger than *Volvox* have not evolved. Larger colonies have more soma and more germ, but the two tissues increase in quite different ways: the increase in the quantity of germ tissue is caused by the greater size of a more or less constant number of germ cells, while the increase in somatic tissue is due to a greater number of smaller cells. This is how the exaggerated division of labour among cells in the larger Volvocales is produced.

The relation between the number and size of gonidia is complex (figure 4). Among undifferentiated forms arising through heterochrony, larger colonies have more and larger cells, all of which are reproductive. Overall size increase thus creates positive correlation between the number and size of germ cells. Among differentiated forms evolving by shifts in allocation, the correlation between the number and size of gonidia is negative, implying a fixed upper limit to the quantity of germ tissue. The tendency for overall size increase to generate positive covariance among fitness components while shifts in the allocation of a fixed quantity of material generate negative covariance is a central principle in the quantitative theory of life histories (Bell & Koufopanou 1986).

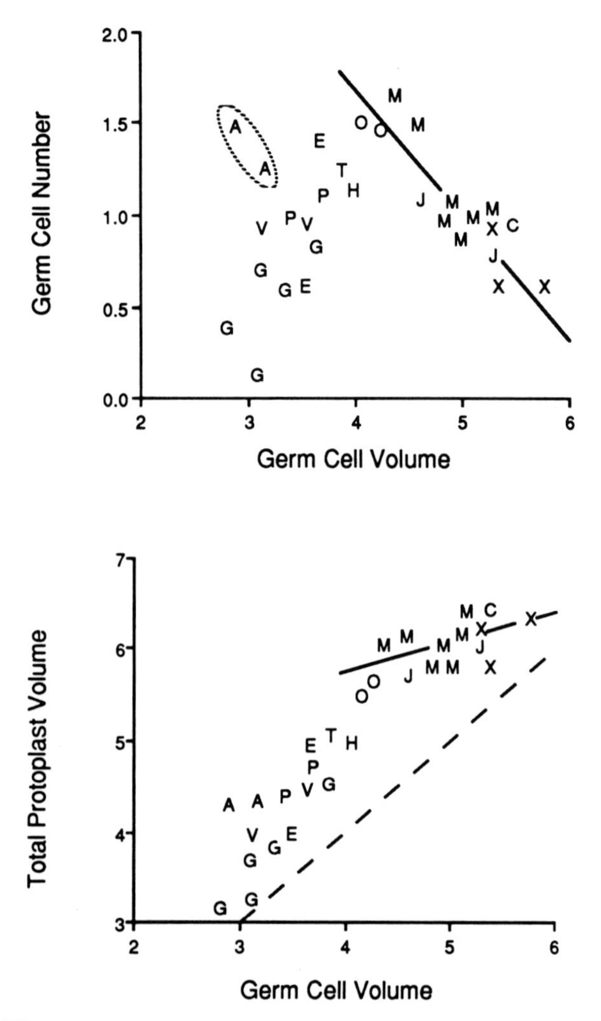

Figure 4. The number and size of germ cells in volvocacean algae of different overall size and degree of differentiation. Letters indicate taxa: A, *Astrephomene*; G, *Gonium*; T, *Platydorina*; V, *Volvulina*; P, *Pandorina*; E, *Eudorina*; P, *Pleodorina*; M, *Volvox*, sect. Merillosphaera; J, *Volvox*, sect. Janetosphaera; C *Volvox*, sect. Copelandosphaera; X *Volvox*, sect. *Euvolvox*. From Koufopanou (1991).

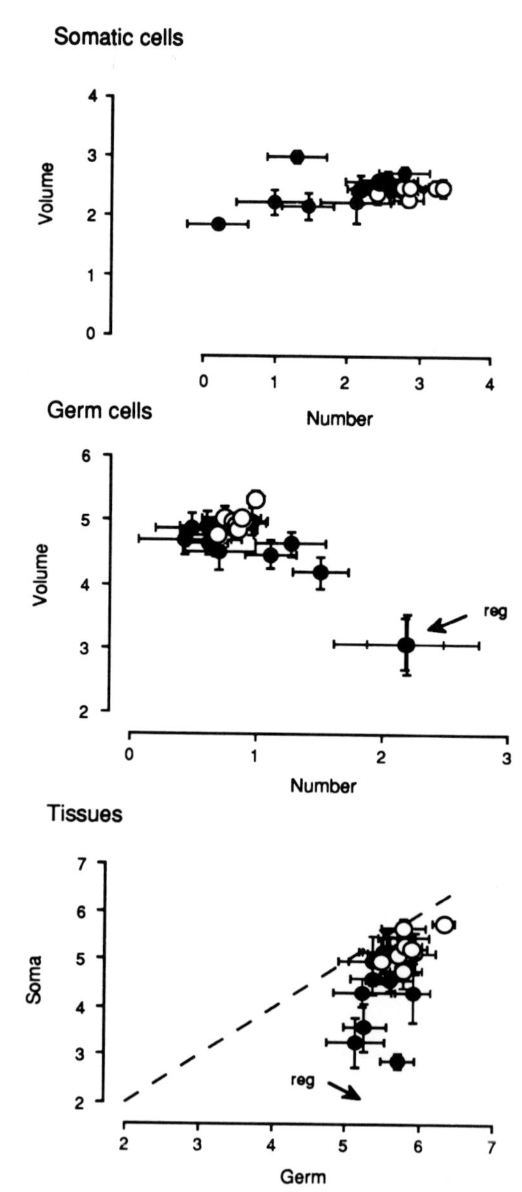

Figure 5. The number and size of cells among mutants of *Volvox carteri*. Wild-type strains are shown as open circles and mutant strains as solid circles; 'reg' is the Regenerator mutant described by Kirk & Harper (1986), in which somatic cells regenerate as gonidia. The bars are ± 2 standard errors. From Koufopanou & Bell (1991).

(*e*) *Developmental mutants of* Volvox

The developmental genetics of *Volvox carteri* are relatively well known and provide an opportunity of looking more closely at the relation between soma and germ. It is possible that somatic differentiation and even multicellularity can be abolished through lesions in only three genes, or closely linked groups of genes (Kirk & Harper 1986). This need not imply that the evolution of differentiated multicellular individuals from unicells requires only three genes, but does make it feasible to obtain by mutation a range of forms which span that found in all the rest of the Volvocales.

We measured the numbers and sizes of cells in a series of such developmental mutants (Koufopanou & Bell 1991 *b*). As expected, the mean values of characters thought to enhance fitness, such as the numbers and sizes of cells, are lower among mutants than among wild-type strains. The total quantity of soma increases steeply with the total quantity of germ, as it does among species; it comprises a highly variable number of cells of rather uniform size. There is a striking negative correlation between the number and size of germ cells, attributable largely to a class of mutants in which somatic cells are able to redifferentiate as gonidia. This again recalls the pattern shown by different species.

The allometry of tissues among species of the Volvocaceae and among mutants of the single species *Volvox carteri* is shown more explicitly in figure 5. The total quantity of somatic tissue varies widely among the mutants, with no relation between the number and size of somatic cells. The negative correlation between number and size found among species might therefore reasonably be attributed to selection. One possible selective agent is the requirement for large numbers of flagella to keep big colonies in motion. The negative correlation between the number and size of germ cells among mutants parallels the pattern among species, suggesting a constraint on the total quantity of germ tissue. The correlation found among species could thus be produced by mutation without the intervention of selection.

4. THE SEXUAL CYCLE

(*a*) *Origins of sexuality*

Sexual cycles in protists are exceedingly diverse, although unfortunately no recent review of the whole group is available. Some large taxa appear to lack sex altogether. Since sex is facultative in most (if not all) protists, it may be that the environmental conditions that induce sex have not yet been discovered in the less well-studied groups. Nevertheless, some very well-studied taxa, such as the euglenids, seem genuinely to lack any conventional sexual process.

A quantitative comparative account of the sexual cycle of Volvocales, complementing the account of the asexual cycle given above, was published by Bell (1985). The discussion here is restricted to experiments with ciliates which emphasize the importance of exogeneous repair in the evolution of the sexual cycle.

(*b*) *Sex and death in ciliates*

The fission-rate of isolate lines of ciliates, maintained as far as possible by single-cell descent, generally declines through time, until the lines eventually become extinct. A sexual episode reverses the decline and restores the previous level of vigour. This phenomenon has recently been discussed at length by Bell (1989). It is thought to be caused by the irreversible accumulation of slightly deleterious mutations at many different loci in asexual lineages (Muller 1964). The ciliate macronucleus is typically highly polyploid and divides amitotically. If a mutation occurs in one of the n copies of a gene, $\frac{1}{n}$ of the descendant macronuclei will bear only the mutant allele, as a result of somatic assortment, and cannot thereafter recover their original non-mutated state. Continued amitosis therefore leads to a monotonic increase in mutational load, causing a steady decline in performance. This can be reversed by the dissolution of the old macronucleus and the formation of a new structure specified by the mitotically replicated micronucleus. However, because the micronuclear genome is not expressed vegetatively, micronuclear mutations are essentially neutral, and will accumulate rapidly, even though they are deleterious when expressed as macromolecular genes. The micronuclear genome can be cleansed of such mutations only by outcrossing followed by recombination. This creates variance in the load of mutations among sexual products and thus enables many deleterious mutations to be eliminated at once through the death of cells bearing heavily loaded genomes, allowing unloaded or lightly loaded genomes to persist, and preventing the irreversible accumulation of mutational damage. The fate of isolate lines of ciliates provides convincing evidence both for the necessity and for the effectiveness of sex as an exogenous repair device. Experimental work has been confined almost exclusively to ciliates, although there is some evidence for similar phenomena in other protists, and even in asexual metazoans.

(*c*) *Sex and natural environments*

The most plausible accounts of short-term selection for the maintenance of sexuality involve the concept that the environment as perceived by the organism continually deteriorates, either through the accumulation of mutations, or through the adaptation of antagonistic organisms, necessitating a continual evolutionary response through sexual outcrossing and meiotic recombination (Jaenike 1978; Hamilton 1980). There is as yet no convincing direct evidence for this theory, although it is consistent with the major ecological correlates of sexuality (Bell 1982), and with the observation that biological control is often more effective against asexual than against sexual pests (Burdon 1987). There is no relevant work on protists. However, it is a commonplace observation that sex is induced by unfavourable conditions of culture. The routine technique for crossing *Chlamydomonas* involves transferring cultures to nitrogen-free medium, which causes the cells to differentiate as gametes and fuse with compatible partners. It would be interesting to know the range of stresses effective in inducing sexuality; whether auxotrophic mutants can be induced sexually by genotype-specific stresses; and whether the rate of recombination as well as the occurrence of sexuality can be environmentally induced. Elucidating the relation of sex and recombination to host–pathogen coevolution awaits the development of a model system using a protistan host.

5. GENERAL DISCUSSION: REPAIR AND REPEATABILITY

A central place in the life cycle is occupied by that single cell or nucleus, which goes by many names in different groups of organisms, but which we have called the spore. Most major taxa of multicellular organisms – the main exceptions are metazoans such as rotifers, nematodes and insects – can reproduce without spores, forming new individuals from large masses of parental tissue by budding or fragmentation. Offspring that are produced vegetatively in this way will inevitably bear a load of somatic mutations. These are transmitted in a non-Mendelian fashion, with some progeny receiving by chance more mutant cells and others fewer, and will therefore show somatic assortment. Any given lineage will therefore tend to become fixed for a succession of somatic mutations. The particular loci that are affected will be different in different lineages, but all will acquire a gradually increasing load of somatic mutations. The process is irreversible because once all cells bear a mutation at a given locus, the original state of the locus cannot be recovered. Using very many cells to form new individuals offers no protection: the assortment of previously acquired mutations will be slowed down, but a greater number of new mutations will arise. Selection will oppose the accumulation of somatic mutations by eliminating heavily loaded individuals, but it will not be very effective because it can act only through the variation among individuals in the mean number of mutations per cell. Lineages which proliferate vegetatively will in the course of time reproduce more slowly, show a greater incidence and frequency of congenital abnormalities of development, and at last become extinct.

This argument introduces the concept of repair, and in particular exogenous repair, as a central organizing principle of life cycles. Exogenous repair is the appropriate exposure of propagules to natural selection. It is related to the concept of 'selection arenas' (Kozlowski & Stearns 1989). Here, we suggest that the spore is an instrument of exogenous repair. When spores are produced, each cell is independently exposed to the action of selection. Genetic variance for effects caused by somatic mutations will be greater among spores than among offspring produced by budding or fission, because the variance of items is greater than the variance of means, and spore production therefore enhances the effect of selection in reducing mutational load. Spores will nevertheless bear a certain number of new mutations, which will increase with the length of time or the number of replications separating spore from spore. Several authors have reported that asexual lineages propagated exclusively through the late offspring of old parents tend to deteriorate and eventually become extinct, whereas those propagated through the early offspring of young mothers flourish (the 'Lansing effect', discussed by Bell (1989), pp. 93–95), and such results, although they remain controversial, may reflect the greater mutational load of older spores. More generally, the details of spore production may be arranged so as to minimize mutational damage. In some organisms the germ line is sequestered early in development and so passes through relatively few replications. This should reduce the necessity for exogenous repair. Consequently, we are led to predict that the overproduction of asexual spores should be greatest among organisms which develop by somatic embryogenesis (multicellular fungi, plants, sponges and a few metazoan phyla), less in organisms with epigenetic development (some molluscs, annelids and echinoderms), and least in organisms with preformistic development (such as gastrotrichs, rotifers, nematodes, tardigrades and vertebrates).

The spore is not the only state to which most or all vegetative cycles are regularly reduced; it is also the switch-point between the vegetative and sexual cycles. What controls this switch? The unequivocal answer is that spores are switched into the sexual cycle when the environment deteriorates, by nutrient depletion, crowding or starvation. The vegetative cycle is the means by which the spore takes advantage of temporarily or locally favourable conditions to grow and reproduce; the sexual cycle is the response of the spore to the eventual failure of these conditions. The reason for being so confident about this statement is that making the environment worse is the routine laboratory procedure for switching spores from vegetative to sexual expression in a very wide variety of protists and other organisms. In a very broad sense, then, the sexual cycle is also a repair mechanism, whose primary function is to overcome the failure of vegetative expression. The sense is so broad, however, that the statement has little substance unless the meaning of 'environment' is made more explicit.

The environment as perceived by an organism might be said to deteriorate in either of two ways. First, the external conditions of growth might change so that current genotypes are poorly adapted, in the sense that fitness would be increased by the appropriate genetic changes. We have already argued that the only way in which the external environment can be expected consistently and continually to deteriorate is through the evolution of antagonistic species. By becoming well-adapted to antagonists, a species causes environmental stress through the selection that it imposes on its antagonists for counter-adaptation, thereby ensuring that further genetic change will be required in the future to maintain adaptation. In some cases, such as the immune system of vertebrates or the antigenic surface proteins of trypanosomes, this change can be achieved endogenously. For the most part, however, it seems possible that some exogenous mechanism such as sexuality is necessary to break up and rearrange currently well-adapted genotypes before they fall victim to their own success. Secondly, if the external conditions of growth were to remain unchanged, organisms might nevertheless perceive them as having got worse if they have received genetic damage such as mutation. Although the production of asexual spores reduces mutational load by exposing individual genotypes to selection, it cannot entirely prevent the irreversible accumulation of deleterious mutations. The sexual cycle arrests this accumulation by generating non-mutant genotypes from mutant gametes through meiotic recombination. For the purposes of this essay, we have drastically simplified the diversity of opinions regarding the effect of sex on adaptation, opinions which are discussed at greater length by Williams (1975), Maynard Smith (1978), Bell (1982, 1989), Hamilton (1980), Kondrashov (1984), and many others. Our main point is that the primary function of the sexual cycle may be to maintain adaptation despite the fact that the external environment or the genotype's ability to deal with the environment are tending continually to deteriorate, and that in this sense it acts as a mechanism of exogenous repair.

The success of the quantitative theory of life histories makes it natural to think that the same principles will be equally successful in the interpretation of life cycles, and in particular to think that the most successful life cycles will be those which have the greatest fitness, in the conventional sense of generating the greatest rate of producing offspring. However, this is not necessarily the case. Life cycles that displace alternative cycles may do so, not only because they are more prolific, but also because they are more precisely repeatable. A theory of this sort has been strongly advocated by Buss (1987), but its intellectual precursors also include C. H. Waddington's concept of epigenetic canalization (Waddington 1957) and R. Dawkins' general theory of replicators (Dawkins 1985). Consider two genes which direct the same sequence of development, leading to the same allocation to germ cells and the same reproductive output; but while one gene invariably directs this life cycle, the other produces the same result only as an average, the actual course of development varying more or less widely among individuals. The former gene is thus associated with a more stable and highly canalized life cycle, and if this life cycle is nearly optimal then its lack of variability will be advantageous. Nevertheless, individuals bearing the latter gene, although their phenotypes may be far from optimal, will produce offspring that will again express the optimal phenotype as an average. Instability of development has more profound consequences when the phenotype of offspring is related to the phenotype of their parent. Any variation in the sequence of development will then increase through time, obliterating the initial sequence completely. If all life cycles were, in a conventional sense, neutral, having the same rate of production of offspring, then the more stable would increase in frequency, not because of their greater reproductive fitness, but because of their greater repeatability. Life cycles will behave in this Markovian manner only if the variation in life cycles among the progeny of a single spore is heritable. Somatic mutations that direct the entry of somatic cells into the germ line will create heritable variation, in the sense that the different life cycle found among the asexual products of a single spore will each tend to be transmitted to the next generation of spores, causing a Markovian disintegration of the original cycle. It is important to distinguish this phenomenon from the effect of 'selfish' somatic mutations in disrupting an orderly and prolific sequence of development. This may often be the case; but the distinctive property of the evolution of life cycles is that heritable stability of development (or the lack of heritable instability) will be favoured on its own account, independently of any effects on the reproductive success of individuals. Stable development will thus be ensured if either the occurrence or the transmission of somatic mutations can be minimized. Buss points out two ways in which this can be achieved: the control of early development by maternal transcripts in the spore cytoplasm prevents the occurrence of expressible somatic mutations, because development is under the control of the nuclear genome of the parent; and the early sequestration of the germ line, besides minimizing germ-line mutations, also reduces the heritability of somatic mutations by denying them access to the germ line. Genes borne by the spore which promote maternal control or germline sequestration will be favoured because they will thereby promote their own precise replication; genes that do not have this property will tend to be replaced by a host of somatic mutations. The phenotypic consequence will be the evolution of highly repeatable life cycles. It is very likely, of course, that, once a stable sequence of development has evolved, selection will favour modifiers that shift the life cycle towards an optimal pattern of allocation which maximizes spore production. Our point is to distinguish between selection for the rate of reproduction and selection for repeatability, and to argue that repeatability, independently of reproductive rate, may be crucial to understanding how life cycles evolve.

This work was supported by an Operating Grant from the Natural Sciences and Engineering Research Council of Canada to G.B.

REFERENCES

Bell, G. 1982 *The masterpiece of nature*. London: Croom Helm; Berkeley: The University of California Press.

Bell, G. 1985 The origin and early evolution of germ cells as illustrated by the Volvocales. In *The origin and evolution of sex* (ed. H. Halvorson & A. Monroy), pp. 221–256. MBL Lectures in Biology, vol. 7. New York: Alan R. Liss.

Bell, G. 1989 *Sex and death in protozoa*. Cambridge University Press.

Bell, G. & Koufopanou, V. 1986 The cost of reproduction. *Oxf. Surv. evol. Biol.* **3**, 83–131

Blueweiss, L., Fox, H., Kudzma, V., Nakashima, D., Peters, R. & Sams, S. 1978 Relationship between body size and some life history parameters. *Oecologia* **37**, 257–272

Bonner, J. T. 1990 *The evolution of complexity*. Princeton University Press.

Brooks, J. L. & Dodson, S. 1965 Predation, body size and composition of the plankton. *Science, Wash.* **150**, 28–35.

Burdon, J. J. 1987 *Diseases and plant population biology*. Cambridge University Press.

Buss, L. W. 1987 *The evolution of individuality*. Princeton University Press.

Dawkins, R. 1985 *The extended phenotype*. Harlow: Longman.

Hamilton, W. D. 1980 Sex versus non-sex versus parasite. *Oikos* **35**, 282–290

Harris, E. 1989 *The Chlamydomonas source book*. New York: Academic Press.

Jaenike, J. 1978 An hypothesis to account for the maintenance of sex within populations. *Evol. Theory* **3**, 191–194.

Kirk, D. L. & Harper, J. F. 1986 Genetic, biochemical and molecular approaches to *Volvox* development and evolution. *Int. Rev. Cytol.* **99**, 217–293

Kondrashov, A. S. 1984 Deleterious mutations as an evolutionary factor. I. The advantage of recombination. *Genet. Res. Camb.* **44**, 199–217

Kozlowski, J. & Stearns, S. C. 1989 Hypotheses for the production of excess zygotes: models of bet-hedging and selective abortion. *Evolution* **43**, 1369–1377

Koufopanou, V. 1991 Development and evolution in the green flagellates. (Submitted.)

Koufopanou, V. & Bell, G. 1991*a* Soma and germ: an experimental approach using *Volvox*. (Submitted.)

Koufopanou, V. & Bell, G. 1991*b* Developmental mutants of *Volvox*: does mutation reflect the patterns of phylogenetic diversity? (Submitted.)

Margulis, L. & Sagan, D. 1986 *Origins of Sex*. New Haven: Yale University Press.

Maynard Smith, J. 1978 *The evolution of sex*. Cambridge University Press.

Muller, H. J. 1964 The relation of recombination to mutational advance. *Mutat. Res.* **1**, 2–9

Waddington, C. H. 1957 *The strategy of the genes*. London: Allen and Unwin.

Williams, G. C. 1975 *Sex and evolution*. Princeton University Press.

Allocation of resources to sex functions in flowering plants

D. CHARLESWORTH AND M. T. MORGAN

Department of Ecology and Evolution, University of Chicago, 5630 S. Ingleside Avenue, Chicago, Illinois 60637, U.S.A.

SUMMARY

The study of allocation of resources offers the possibility of understanding the pressures of natural selection on reproductive functions. In allocation studies, theoretical predictions are generated and the assumptions as well as the predictions can be tested in the field. Here, we review some of the theoretical models, and discuss how much biological reality can be included in them, and what factors have been left out. We also review the empirical data that have been generated as tests of this body of theory. There are many problems associated with estimating reproductive resources, and also with testing how allocation of these resources affects reproductive and other components of fitness, and we assess how important these may be in allowing empirical results to be interpreted. Finally, we discuss the relevance of resource allocation patterns to the evolution of unisexual flowers, both at the level of individual plants (monoecy, andro- and gynomonoecy) and at the population level (dioecy).

1. INTRODUCTION

An understanding of the patterns of allocation of resources to different functions and structures in living organisms implies knowledge of the way natural selection acts on these functions. Therefore an important part of the study of the evolution of plant breeding systems concerns allocation to the structures and functions involved in the various stages of the reproductive process. Theoretical models of allocation to these structures enable one to incorporate interactions of plants and their environments into theories of breeding system evolution, in a disciplined and organized way. This approach complements research on the genetic advantages and disadvantages of different plant breeding systems.

When one constructs a model of allocation of reproductive resources, one tries to include all the functions and structures that can affect the numbers of progeny produced. In hermaphrodite organisms, such as most flowering plant species, this means including reproduction via both male and female functions, and including all stages of the reproductive process (Lloyd 1975; B. Charlesworth & D. Charlesworth 1978). For example, if seedlings compete with one another, the increase in the number of surviving progeny with number of seeds produced would be sharply limited, and seed number would be a poor indicator of progeny production through seeds (Lloyd 1979). Therefore, to be useful either as a conceptual framework for thinking about the costs and benefits of different reproductive functions, or to be analysed quantitatively, models of allocation must involve considerable biological realism, and include many variables with important effects on the evolution of allocation patterns. The inclusion of this degree of realism, however, causes various problems. Complex models, which take most of the important factors into account, have so many parameters that they are difficult to analyse. Alternatively, one can focus on certain effects, and construct simplified models that omit other factors. In either case, the predictions of these models are often hard to test, because species seldom differ in just the parameter of interest. For example, the prediction that plants with a certain type of pollinator should allocate more to attractive structures than those pollinated by a different type of pollinator appears to be simple and testable, until one realizes that the difference in pollination is likely to be associated with other differences, such as in the degree of self pollination, or the time of year when pollination occurs. There are thus difficult problems to be solved in the testing of allocation theories.

In this review, we start by outlining the way in which models of allocation to sex functions in hermaphrodite plant populations can be constructed and analysed. After describing simple allocation models, we review recent models with greater biological realism (i.e. more parameters). We then review the evidence that there is a single resource pool for the different reproductive functions, and the problems involved in measuring allocations and in testing the model's predictions. Finally we discuss the insight that allocation studies can give into the evolution of unisexuality.

2. THEORETICAL MODELS OF SEX ALLOCATION IN COSEXUAL PLANTS: MODELS AND METHODS OF ANALYSIS

(*a*) *Annual life cycle*

Theoretical models of sex allocation assume that the resources available for reproduction are fixed and can be devoted in varying amounts to different functions (reviewed by Lloyd 1987*a*, *b*). The reproductive fitness of a cosexual organism is the sum of contributions due to male and female reproduction

$$w_h = w_f + w_m$$

where w_f is the female and w_m the male contribution. Assuming some genetic variation in allocation patterns due to variation at nuclear loci whose alleles act additively and have no selective effects apart from those on allocation, the relative fitnesses of different phenotypes are proportional to the numbers of gametes transmitted to progeny at the same stage in the next generation (Lloyd 1975, 1977; Charlesworth & Charlesworth 1978; Gregorius & Ross 1981). These fitnesses depend on the amounts of resources allocated, the effects on the numbers of ovules and pollen grains produced, and their success in the fertilization process and the subsequent stages of seed maturation and seedling establishment. Mating systems with both self- and cross-pollination can be modelled, as well as allocation to structures for pollinator attraction (including petals and nectar). These consume resources that could otherwise be used to produce more ovules or pollen.

The female contribution to fitness depends on the numbers of seeds and seedlings produced. Seed output depends on the proportion, F, of reproductive resources allocated to ovules, and on the allocation to attraction of pollinators, A. The probability of selfing will depend on details of the flower size and the times of maturity of the anthers and stigma, and also on the probability of pollinator visitation (Lloyd 1979). Models of allocation can include various possibilities for selfing, which may occur before pollinators visit (prior selfing), or by the agency of pollinators carrying a mixture of self and outcross pollen (competing selfing), or by self-pollination of unvisited flowers (delayed selfing). In general, therefore, we should consider selfing rates to be functions of A, $S(A)$ say (Lloyd 1987*c*). These factors can be put together (for the case when all ovules are fertilized either by prior or competing selfing, or by outcross pollen) to get an expression for the number of gametes a plant contributes through seed production:

$$O(F)\{1-S(A)+2S(A)(1-\delta)\}, \qquad (1)$$

where $O(F)$ is the number of ovules and δ is the inbreeding depression (the reduction in fitness of selfed, relative to outcrossed, progeny).

Finally, the female contribution to fitness depends on the number of fertilized ovules matured into seeds. Two extreme situations are biologically realistic. In some plants, fruits mature after flowering and may therefore draw on a different resource pool from that available at flowering time. Because allocation models assume that there is a fixed total amount of resource for reproduction, a structure or function should be included only if changing it affects the expression of other functions. This is most likely for functions that occur simultaneously. In the case when flowering and fruiting have separate resource pools, F will be the allocation to female functions (ovaries and ovules) at the time of flowering, and the female component of fitness will have an upper limit dependent on the second resource pool (Charlesworth & Charlesworth 1987). The opposite extreme is a single resource pool for reproduction, limiting both flowering and fruiting (Lloyd 1987*c*). F is then the proportion of this amount that is used up in both stages of female reproduction. In many plants, fruits start to mature while flowers are still developing, so the reality clearly lies between these extremes. The appropriate assumptions for modelling, and for estimating the parameter F empirically, must therefore be decided according to what is most appropriate for the particular plant of interest, thus reducing the generality of the results.

Now consider male reproduction. This is necessarily competitive, since the effect of increasing pollen output on the numbers of ovules fertilized depends on the amount of pollen produced by other individuals in the population. The contribution to fitness through male function will be a function of M, the proportion of resources allocated to anthers and pollen (which equals $(1-A-F)$, because of the assumption of a fixed total amount of resource). It will also depend on visits from pollinators: the contribution to the pollen pool for outcrossing will be a function of both M and A, say $P(M, A)$. The contribution to fitness of a given phenotype through male function is therefore given by the number of ovules available to pollen in the pollen pool (which is simply the number of ovules that are not selfed), multiplied by the ratio of its pollen production to the average pollen output. We can write this fitness contribution of a particular phenotype, j, in a population with any number of allocation phenotypes designated by subscripts i and having frequencies z_i, as:

$$w_m = \{\sum_i z_i O(F_i)[1-S(A_i)]\}\frac{P(M_j, A_j)}{\sum_i z_i P(M_i, A_i)}. \qquad (2)$$

Analysis of the evolution of allocation patterns based on this type of fitness equation can be achieved by searching for a set of values of allocation parameters such that, if a population existed with that set, any other slightly different values would produce a lower fitness, i.e. the fitness of the phenotype expressed by the population is higher than that of an invading rare type, so the new type will not increase in frequency. To find this evolutionarily stable strategy (ESS, see Maynard Smith 1982; Lloyd 1987*a*, *b*), the fitness equation for a particular phenotype is written as $w(\boldsymbol{a}|\boldsymbol{a}^*)$, where $\boldsymbol{a}$ refers to a vector of allocations defining a phenotype, and $\boldsymbol{a}^*$ is the vector for the population. The ESS can be found by solving for the values of the allocation parameters at which the partial derivatives of fitness with respect to the components of $\boldsymbol{a}$ are zero, and the second derivatives negative, with the derivatives evaluated with the allocation set having the values of $\boldsymbol{a}^*$ (Charlesworth & Charlesworth 1981).

Other methods of analysis derived from this fundamental method may be technically easier to use. For example, Charnov (1982) discussed the 'product rule', which applies to many allocation problems, and states that the ESS allocation parameters maximize the product of the male and female contributions to fitness.

Lloyd (1987*a*) proved a marginal value theorem stating that at the ESS allocation set the marginal gains for the functions, weighted by their reward values, are equal. Lloyd (1987*a*, *c*) has used this to solve for allocation problems in which one resource is partitioned between multiple functions. Sometimes analytical approaches may be impossible, and graphical methods may be used (see, for example, Charlesworth 1989). Charlesworth (1990) showed that the results of ESS analyses give approximately the same results as those obtained from quantitative genetic analysis, for weakly selected characters.

We must next describe how the allocations translate into the fertilities that appear in the fitness expressions. It is biologically reasonable to think that these relations should be increasing functions, but that they would rarely be linear (Charnov 1979). At least one of the male and female gain curves (and often both) will usually be saturating. This will be the case when the gain through either sex is limited by factors other than the amount of resources allocated. For example, pollen contributions to the outcrossing pollen pool will be limited by the numbers of pollinators and their capacity to carry pollen (Charnov 1979). There would therefore tend to be a diminishing returns, or saturating gain, effect for increasing pollen production. This shape of curve is also reasonable for female fertility, because seed production is often limited by resources for fruit maturation, and thus has a separate limit from that imposed by resource limits on flower and ovule production (Willson 1983). We return below to some interesting cases when one or the other curve might be accelerating.

To incorporate the shapes of the gain curves into the models, the allocations can be raised to suitable exponents. For example, to model the saturating gain for male fertility, one can raise the pollen output to a power less than 1. This method has some convenient properties for solving for allocation patterns (Charnov & Bull 1986), and is also reasonable in biological terms, although S-shaped curves would probably be more realistic (Frank 1987). Charnov & Bull (1986) and Lloyd (1989) showed that for allocations between multiple functions with this type of model in populations with no inbreeding, the ESS allocations for outcrossing populations are in the ratios of the exponents. In other words, a function with severely limited gain will tend to be allocated lower amounts of resources. Furthermore, the relative amounts of resource allocated to two functions can be deduced without considering other functions that draw on the same resource pool (Lloyd 1989).

These models can be extended to include allocations to one function that may affect the success of another, for example if pollen attracts pollinators so that greater pollen output affects the chance of fertilization of the ovules, as well as the male fertility (Lloyd 1987*c*; Charlesworth & Charlesworth 1987). The selfing rate may also be a function of the male allocation (Charlesworth & Charlesworth 1981). It is also possible to study models in which functions draw on more than a single resource pool. McGinley & Charnov (1989) have done this for the problem of seed size and number. The possibility of different environments, in which the gain curves differ, has also been analysed (Charnov & Bull 1985). We will not attempt to describe these elaborations in detail here.

Another type of sex allocation model explicitly includes costs per item, such as ovules, pollen and flowers (for examples, see Cohen & Dukas 1990; Spalik 1990). Assuming that all reproductive functions, including fruiting, draw on one resource pool, so that increased allocation to flowering-time functions, such as flower numbers or attraction, reduce seed output, these models can give results about ratios of fruits to flowers (Morgan 1991), which are not possible with the kind of models described above. Constancy of the costs of the unit reproductive structures is also usually assumed (but see Schoen & Dubuc 1990).

(*b*) *Perennial life cycles*

In perennial plants with reproduction at more than one age, allocation to reproduction may affect the probability of survival to the next age class, and this can affect the sex allocation that will evolve (D. Charlesworth 1984; Charnov 1988; Kakehashi & Harada 1987). The difficulty in analysing such models is the number of parameters involved. These include the allocations to male and female functions of different age or size classes, and the effects of allocation on the transitions to the next stages. When the survival probabilities are independent of the allocations, the results are similar to those for annual life cycles (Charnov 1988). When allocation to female functions reduces survival more than male reproduction, but the effects are independent of age, allocations shift towards more male function (Tuljapurkar 1990).

These results assume that the life history is fixed, and do not allow for the joint evolution of sex allocation together with the onset and frequency of reproduction. Given the difficulty of producing general theories of life-history evolution (reviewed by B. Charlesworth 1984), it is not surprising that this even more complex problem has not been solved. Charlesworth (1990) describes a method based on quantitative genetics which is useful for this type of study of multiple allocation strategies. He used it to find ESS life histories assuming that at each age class resources may be allocated either to reproduction or to survival. Inclusion of different effects on survival of allocation to different sex functions would probably change the life histories predicted to evolve. Male reproduction probably often affects survival less than fruiting, and in many perennial plants male reproduction starts at an earlier age (or at smaller size) than female reproduction (Godley 1964; Meagher 1982). Inclusion of these sex differences into models of sex change with size has been strikingly successful in predicting the behaviour of such plants (Policansky 1981; reviewed by Charnov 1982). Investment per offspring may also evolve to be higher in harsh environments (Kawano 1981). If this happens, it will be important to take it into account in the theory. It would certainly be valuable to have more data on between-population variation within species, especially species occupying a range of habitats.

3. EVIDENCE FOR LIMITED POOLS OF RESOURCES

A major assumption, explicitly underlying theoretical models of sex allocation, is that there is a fixed amount of some resource which is available for reproductive functions alone, and which can be split up in various possible ways among the different reproductive functions (see Partridge & Sibly, this symposium). This assumption is applied strictly only in models that assume an annual life cycle. In perennials, there is also the possibility for unused resources from one reproductive episode to be used for growth and to increase survival to the next breeding season. The idea of limiting pools of resources is an old one and seems intuitively reasonable, and is a basic assumption of models of life-history evolution (see Partridge & Sibley, this symposium). Darwin considered this to be a fundamental property of biological systems, which he referred to as 'compensation' and viewed as producing an increase in the amount or size of one structure when another is reduced, for example an increase in female fertility in male-sterile plants (see, for example, the discussion in Darwin 1877, p. 280; Stauffer 1975). In the recent literature, this type of effect is usually referred to as a 'trade-off'.

Strangely, the empirical support for this principle is not very extensive (Antonovics 1980). It is important to distinguish between the concept of gender specialization in terms of allocation of more resources to one sex function than to the other, and the fact that the dependence on the fertility of other individuals in reproduction by outcrossing causes an inverse relation between the realized male and female reproductive success of different individual plants in a population (Robbins & Travis 1986). Here, we are concerned with differences in the resources devoted to different functions, and the extent to which those functions can be developed. For example, do plants with bigger than average petals tend to have fewer ovules, or smaller seeds?

A difficulty in these types of studies is that several factors that affect overall plant 'quality' can induce positive correlations between such phenotypes as the number of flowers and their size, with low quality plants having low values of the different measured characters. These factors include age effects, genetic differences between populations, different degrees of inbreeding (if the characters show inbreeding depression so that some plants will tend to have low values for independent characters), and environmental heterogeneity (B. Charlesworth 1984, 1990). Thus, although negative phenotypic correlations certainly suggest trade-offs, they are unlikely to be found in real organisms. This is a well-known problem in studying trade-offs (Bell & Koufopanou 1986).

To date, most published studies are at the purely phenotypic level (see, for example, Stanton & Preston, 1988; MacNair & Cumbes 1990). The problems can be minimized by correcting for size differences, which should be good indicators of environmental quality and of levels of inbreeding, but this is not always done. It would be better to include estimation of genetic correlations, which should uncover negative relations, if these exist. A practical alternative for the study of some kinds of trade-offs involves manipulation experiments. Horvitz & Schemske (1988) allowed randomly chosen individuals of a tropical herb species either to set fruit by open pollination, or supplemented pollination so that more fruits than normal developed. This treatment difference had no detectable effect on survival or on flowering in the subsequent year, suggesting that there was no trade-off between fruiting and survival in the species studied.

The evidence for trade-offs in reproductive characters is well reviewed by Goldman & Willson (1986). In a few perennial species, negative correlations between heavy fruit production and growth have been convincingly shown (Eis *et al.* 1965; Rohmeder 1967; Jing & Coley 1990). In some dioecious species female plants grow more slowly than males (Meagher 1982; Sakai & Sharik 1988), but the differences are not always in this direction (Grant & Mitton 1979). Females also often flower less frequently than males (see Ornduff 1987) and may experience greater mortality, although there have been few studies in natural populations (Savage & Ashton 1983; Meagher 1982) and results in the opposite direction have also been found (Dawson *et al.* 1990). Meagher (1982) showed that the sex differences in size occur only after reproduction has occurred, which is evidence that they are due to the energetic burden of fruiting, as should be the case if they are really reflections of allocation differences, but most studies do not include such evidence.

Another type of evidence for resource limitation of one function when another related function is exercised at a high level comes from some well-known negative size-number relations. Seeds from fruits with many seeds are usually smaller than those from few-seeded fruits, and plants with many fruits tend to have smaller ones (Snow & Snow 1988). Similar differences have been found between species with different seed sizes (Kawano 1981), and in pollen sizes between males and hermaphrodites of two subdioecious species (with males having larger pollen (McKone & Webb 1988)). These relations seem likely to be the results of allocation differences. We will not discuss size-number relations further in this review, as they have recently been reviewed by Lloyd (1989). Trade-offs between reproductive structures have rarely been studied (Stanton & Preston 1988). Evidence that producing more ovules entails a cost in terms of lower pollen output, or that having large petals tends to lead to fewer ovules, is badly needed.

4. MEASURES OF ALLOCATION

Measures of allocation are involved in testing both the assumptions of allocation models and their predictions, but there are some difficulties with these estimations. The problems of using dry weight as an estimator of allocation to different functions have been emphasized (Lovett-Doust & Cavers 1982; Goldman & Willson 1986; McGinley & Charnov 1989). Because of differences in the composition of different structures (e.g. in their nitrogen and phosphorus contents), dry weights do not correctly estimate total allocations of resources to different structures (Lovett-Doust & Harper 1980). However, the problems may not be as severe as has been assumed. An allocation model specified in terms of dry weights could be tested if one had estimates of the gain curves in a particular population, based on individuals differing in the biomass of their reproductive structures. The allocation pattern for the population could then be compared with the model's predictions (in the same units). It may also be reasonable to assume that the relations between dry weight and allocations to different functions are similar in different individuals of the same species, and in closely related species, and to use dry weights to compare patterns of allocation to different functions. This assumption could be tested.

It is important to note here that the type of allocation model described above assumes that there is variation in the proportions of reproductive resources allocated to different functions, but that the total quantity of resources is fixed. In reality, plants will differ in the total amount of resources available for reproduction. This introduces a problem for studies of gain curves. The curves estimated in field studies represent the relations between the absolute amounts of resources devoted to different functions and the contributions to fitness, whereas the theoretical allocation models are expressed in terms of the proportions of the total resources. As for studies of trade-offs discussed above, it may be possible to take this into account by scaling by some measure of plant quality, such as plant size.

It may thus often be best to estimate allocations by manipulation experiments in which differences in plant quality that may affect total quantities of resources available for reproduction can be avoided, or else explicitly included in the experimental design. Silvertown (1987) estimated the cost of male function in units of plant growth, in a manoecious cucumber species, by measuring the growth of plants whose male flower buds were removed, compared with intact plants (and also by comparing the effects on growth of removing just female, or both male and female flowers). The cost of female flowers, in the same units, was estimated by removal of female flowers (and also by comparing plants whose male flowers were removed with plants with all flowers removed). Whether the growth differences were measured as number of nodes formed, or as dry weights, the results were similar: plants with male flowers removed grew by amounts similar to plants with no flowers removed, and plants with female flowers removed grew no bigger than plants with both male and female buds removed. These consistent results suggest that the cost of male flowers is slight, even though their dry weights were not, and that the cost of female functions (flowers and fruits) was the major cost of reproduction. This is not surprising, because this is a plant with large fruits (owing in part to human selection for large size). The experiment therefore confirms other evidence that, when allocation to fruit maturation is included in female allocation, female functions consume more than male functions.

Populations polymorphic for different sex morphs may offer an opportunity to estimate allocations to male and female functions. Gynodioecious species have male-sterile individuals in populations. Preventing pollination in female or hermaphrodite plants leads to an increased number of flowers. If there is a cost of male functions, the cost per flower will be less for female than hermaphrodite flowers. If the resources not used by fruits can be used to make more flowers, female plants prevented from fruiting should make more flowers than hermaphrodites. In an experiment with paired sibling plants of *Silene vulgaris*, this did indeed happen, so that in this species there appears to be a measurable cost of male structures (D. Charlesworth, unpublished observations).

Ross & Gregorius (1983) used the principle of trade-offs to estimate proportions of resources allocated to different functions, for populations in which individuals with differing allocations occur. For each phenotype in such a population, the cost per seed (r_f) multiplied by the number of seeds (ϕ) added to the cost per pollen grain (r_m) multiplied by the number of pollen grains (μ) should equal the total amount of reproductive resources (assumed to be fixed). Thus, for two phenotypic classes of individuals, there are two equations which could be solved for the two unknown quantities, allocation to seeds ($R = r_f \phi$) and to pollen:

$$\left.\begin{array}{l} r_f \phi_1 + r_m \mu_1 = 1 \\ r_f \phi_2 + r_m \mu_2 = 1 \end{array}\right\} \quad (3)$$

For *Leavenworthia crassa*, using data of Lloyd (1965) from two flower colour morphs, the estimated allocations to female functions were 0.57 and 0.65. Two *Lupinus nanus* flower colour forms studied by Horovitz & Harding (1972) gave estimates of 0.67 and 0.76. These results are limited by the assumption that the total amount of resources is the same for all plants.

In gynodioecious populations there is known to be genetic variation at loci affecting sex functions (this may also exist in cosexual species, but the evidence for genetic variation in sex functions in these is tenuous). Atlan *et al.* (1990) used between-family variation of plants from populations having different frequencies of female plants, to detect negative correlations for the numbers of germinable seeds and pollen grains per flower, scaled to correct for differences in plant size. Their results suggest trade-offs between pollen and seed output. In addition, the data provide estimates of the cost per seed, in units of numbers of pollen grains. The values for the two sets of plants studied were 1330 and 4770, respectively. Alternatively, a plant producing no pollen could produce 2.4–2.6 times as many seeds as one that did produce pollen, a value close to the observed difference between females and hermaphrodites in this species.

5. PREDICTIONS OF ALLOCATION MODELS

In reviewing the results of allocation models, we will focus attention on the patterns they predict, rather than on quantitative predictions of amounts of resources that should be allocated. Although the theoretical models generate quantitative predictions of allocation amounts, given assumptions about the values of the parameters involved, there is in reality little possibility for testing them quantitatively. Firstly, as just discussed, it is difficult to measure allocations to different functions. In addition, no model includes all possible parameters of importance, and it is impossible to measure all parameters. It is, however, possible to predict patterns of allocation differences. Comparative data may therefore be adequate for testing the predictions of sex alloction theory, because if one has enough data the values of the parameters of interest should often be uncorrelated with the values of other parameters (Queller 1984).

Many results of allocation theory were reviewed by Charnov (1982), who stressed the similarity to sex-ratio theory for dioecious species. The chief results that have been obtained may be summarized as follows.

1. Allocations to male and female functions will be unequal when the gain in the contribution to fitness differs for equal proportionate increases in allocation to the two different sex functions, i.e. the gain curves are nonlinear (Charnov 1979). Despite the difficulty in measuring allocations, it appears highly likely that this prediction is fulfilled in plants.

2. With selfing, plants are expected to allocate less to male functions, because with fewer ovules available for the outcrossing pollen pool to fertilize, resources expended on pollen gain less in terms of fitness (Charlesworth & Charlesworth 1981). It is well documented that selfing species produce small amounts of pollen (Cruden 1977; Schoen 1982; Preston 1986; Vasek & Weng 1988; Cumaraswamy & Bawa 1989).

3. Allocation to attractive structures may be high in outcrossers, but is expected to be low in selfing populations (Charlesworth & Charlesworth 1987; Lloyd 1987*c*), in agreement with many observations of small flower sizes of self fertilizing plants (Ornduff 1969; Schoen 1982; Cruden & Lyon 1985; Ritland & Ritland 1989). When, however, both cross- and self-fertilization are mediated by pollinator visits (e.g. when selfing occurs by pollination between flowers on the same plant), attraction can still be important in selfers (Lloyd 1987*c*). Allocation to attractive structures should also be high when pollinators are limiting and competition occurs for insect or animal visits. This may be hard to test, because this situation can also select for smaller flower size and autogamous self-fertilization (reproductive assurance), but the case of alpine and arctic flowers, which appear highly attractive for pollinators (Kevan 1972), may be relevant here.

4. In wind-pollinated plants, loss of a large fraction of the pollen should produce a roughly linear male gain curve (with a low ratio of number of ovules fertilized per pollen grain produced, so that increasing pollen output would increase male success proportionately, but see Burd and Allen (1988)). In contrast, the limited pollen-carrying capacity of pollinators should cause a decelerating gain with pollen production, in animal-pollinated species. Allocation to pollen is thus expected to be higher in wind-pollinated than animal-pollinated species. If pollen of wind-pollinated species has similar, or even considerably smaller, costs per grain compared with pollen of animal-pollinated species, high pollen output is expected. Wind-pollinated plants are indeed known to produce 'an enormous quantity of pollen' (Darwin, 1877, p. 281). The sole study of the male gain curve in a wind-pollinated plant suggests an initially accelerating gain, possibly levelling off at very high male allocation (Schoen & Stewart 1986). For animal-pollinated plants, there is so far only slight evidence for decelerating gain curves (Thomson & Thomson 1989; Young & Stanton 1990).

Some other predictions about allocation patterns can be made by considering the shapes of male gain curves. Queller (1984) suggested that species with pollinia, or with clumped pollen, should have low pollen : ovule ratios. The main reason for this prediction is that when all the pollen from a flower is transferred to a single recipient there will be a more sharply limited gain curve for increased pollen amounts per flower.

Recently, several new models have been studied. Because increased pollen production may increase selfing we should not treat selfing rates as fixed parameters of our allocation models, but should allow them to vary in the same way as other features of the breeding system. There is a small amount of empirical support for an effect of male allocation on selfing rates (Schoen *et al.* 1986), but such data are difficult to obtain. Such differences help to maintain intermediate selfing rates in populations, rather than their evolving to the extremes of total outcrossing or total selfing (D. Charlesworth & B. Charlesworth 1978, 1981; Gregorius 1982). As many plant populations may have intermediate selfing rates (reviewed by Baker 1959; but see Schemske & Lande 1985; Barrett & Eckert 1990), this possibility is worth serious attention. Holsinger (1991) recently suggested that pollen removal from flowers during pollinator visits that cause selfing may decrease the contribution to the outcrossing pollen pool. Holsinger showed that this 'pollen discounting' (which has generally been thought to be an unimportant phenomenon, see Piper & Charlesworth (1986)) tends to maintain intermediate selfing rates as evolutionarily stable states. It may be hard to imagine how removal of the small amounts of pollen that are needed for self fertilization of flowers could significantly reduce male fertility in outcrossing, but it appears quite realistic when pollinators must align themselves accurately in relation to the flower, and pollen is placed on a specific part of their body. Then if that part comes in contact with the flower's stigma before the pollinator leaves and causes selfing, it must also cause a substantial reduction in the amount of pollen remaining to be transported to other stigmas. This interesting mode should certainly stimulate more field work on pollen discounting.

Another situation in which allocation to the two sex functions might be expected to differ is when there are two populations with pollen flow mostly in one direction, so that the pollen in one sub-population is diluted by pollen coming into it from the other. One might expect the population receiving the most pollen from the other sources to evolve lower male allocation and greater femaleness. However, Kirkpatrick & Bull (1987) showed that this is incorrect. Because the female contributions to fitness are not affected, assuming that pollen supply does not limit female fertility, the evolutionary outcome depends only on the male contributions. Equation (2) above shows that it is not absolute, but relative, male fertilities that matter, so that both sub-populations will be subject to the same selection on male function.

In a plant whose flowers develop in a sequence, the male and female contributions to fitness are each due to the sum of contributions from flowers at each of the stages in the sequence, and these contributions may differ so that pollen from different flower stages has different opportunities for siring offspring. If the stages differ in the ratios of numbers of ovules available in potential mates to the amount of pollen produced by the competitors for the pollen pool for those ovules, there can be selection for differences in sex allocation in flowers of the different stages (Brunet 1991). Various factors could produce differences in the ratios. In protandrous plants, where the anthers of each flower mature before the pistils, flowers produced early will have a low ratio of available ovules to pollen competing for them, compared with later flowers. This is expected to select for female-biased allocation patterns in the early flowers (Darwin 1877, p. 283; Pellmyr 1987), and the models of Brunet confirm that this selection pressure indeed operates. This interesting result remains true with partial selfing and when, as in the models discussed above, the male and female fertilities are nonlinear functions of the allocations. It is testable empirically, and does indeed appear to occur (Thomson 1989; Brunet 1991; Spalik & Woodell 1991). Moreover, the tests involve comparisons of flowers on inflorescences of individual plants. Therefore the different stages are comparable in the sense that there is no possibility that comparisons will be vitiated by differences in inbreeding coefficients between parental plants, or by environmental differences.

6. SEX ALLOCATION AND THE ADVANTAGES OF UNISEXUALITY

One of the most interesting aspects of the study of allocation to reproductive functions is the light this may shed on the evolution of unisexuality. Consideration of expected allocation patterns should give insight into the ecological conditions most favourable for the evolution of unisexuality. These considerations have led several workers to attempt quantitative estimates of allocation parameters (see, for example, Silvertown 1987), and have motivated study of the shapes of the gain curves for male and female fertility in plants. Even without quantitative estimates of the amounts or proportions of resources allocated to reproductive functions, we might still be able to deduce what situations would make the loss of one sex function most probable, and thus generate testable predictions. These might concern the evolution of unisexual flowers within plants, i.e. the evolution of monoecy or of andro- or gynomonoecy, or the evolution of entirely unisexual individuals, i.e. the evolution of gyno- or androdioecy and dioecy.

It has sometimes been argued that when a cosexual plant has a high allocation to one sex function there should be the possibility for large gains to unisexual forms through the other sex (Givnish 1980). For example, if fruits are expensive, males could acquire large amounts of resources for increased pollen production. An accelerating gain for allocation to pollen might then permit the invasion of cosexual populations by males (see Bawa 1980). But this does not take into account that a high allocation to female function implies that the male gain is not strongly accelerating. It is therefore preferable not to use arguments of this kind, but to consider separately the invasion by female and male phenotypes. By finding the ESS allocation pattern of a cosexual phenotype, one can ask whether the increase in fitness of male or female phenotypes would be sufficient to allow them to invade the cosexual ESS population (Charlesworth & Charlesworth 1981; Charlesworth 1984). Such models can include the possibility of partial self fertilization in cosexual plants, as the avoidance of inbreeding is another factor that may be important in breeding-system evolution (B. Charlesworth & D. Charlesworth 1978).

(*a*) *Does evolution of unisexuality involve re-allocation of resources from one sex function to benefit the other?*

It seems likely that re-allocation of resources plays a major role in the evolution of dioecy. Darwin (1877, p. 279) suggests, for the evolution of dioecy: '...if a species were subjected to unfavorable conditions from severe competition with other plants, or from any other cause, the production of the male and female elements and the maturation of the ovules by the same individual, might prove too great a strain on its powers, and the separation of the sexes might then prove highly beneficial'.

Is there evidence for such processes in species that are evolving towards dioecy, such as sub-dioecious and gynodioecious species? We have considered gynodioecy above. Even when outcrossing is performed by hand, females of *Thymus vulgaris* produce more fruits per flower and more or larger seeds than hermaphrodites (Assouad *et al.* 1978). Such differences cannot be the result of inbreeding effects on the seeds of hermaphrodites. These data therefore suggest that hermaphrodites allocate more to male than to female reproduction. Further evidence comes from sub-dioecious plants. In these, females are generally quite constant in their sex phenotype, but the plants with male function often range from hermaphrodites with substantial seed set to individuals that are functionally nearly completely male (Westergaard 1958; Webb 1979). Once gynodioecy is present in a population, there will be selection for the cosexual morph to specialize in greater male function (B. Charlesworth & D. Charlesworth 1978; Charlesworth 1989). Unless there are trade-offs between male and female function, however, (e.g. if these functions had separate resource pools) increased male function need not involve loss of fruit-producing ability. The loss of female functions which occurs in the evolution of the males of sub-dioecious species is therefore evidence for a common resource pool for male and female functions. It would be very interesting to have more data on sub-dioecious populations, particularly studies of the genetic correlation between male and female functions among the plants having male function (Sakai & Weller 1991).

Femaleness may sometimes evolve without major re-allocation of resources, at least at the level of flowers. In several dioecious species, hermaphrodites have large anthers containing pollen, giving the appearance of androdioecy, but the pollen is either non-functional, or else the anthers do not dehisce during the time that stigmas are receptive (reviewed by Charlesworth 1984; see also Liston *et al.* 1990; Kawakubo 1990; Schlessman *et al.* 1990). However, females could gain extra resources if their flowers produced much less pollen, as in *Solanum* (Anderson 1979) or fewer stamens as in *Eugenia* (van Wyk & Dedekind 1985), or if there were many fewer flowers (Anderson & Symon 1989). Similar data come from species that are dioecious but in which flowers of the two sexes are morphologically similar ('cryptic dioecy'). There is usually some degree of reduction in the gynoecia of males, sometimes merely absence or reduction of the ovules, and the anthers of the females may differ only slightly from those of males (Mayer 1990; Kevan *et al.* 1990). Some re-allocation has thus probably occurred. Males may also have more flowers than females (Kevan *et al.* 1990), as is also common in many dioecious species (reviewed by Lloyd & Webb 1977). It would be interesting to have quantitative measures of the allocation to reproductive structures in the two sexes of cryptically dioecious species.

(*b*) *Theoretical results*

Charnov *et al.* (1976) first showed that the shape of the curve relating male and female contributions to fitness (which in turn depends on the relations between allocation to the two sex functions and these two contributions) can determine the stability of dioecious populations to invasion by hermaphrodites or other cosexes that can reproduce as both male and female. When the relation is bowed outwards, hermaphroditism can invade, whereas accelerating gains for the contribution of one sex to fitness may explain the evolution or maintenance of dioecy (Charnov 1982; Lloyd 1982). Although invasion by cosexes may be prevented by the presence of inbreeding depression, even when the shape of the male–female curve is decelerating (see below), this curve may nevertheless help us gain understanding of when dioecy is most likely to be maintained (Charlesworth & Charlesworth 1981). Many treatments of the evolution of dioecy concentrate on verbal arguments suggesting how various factors will affect the male–female gain curve (Thomson & Brunet 1990).

For example, the female gain curve depends strongly on the dispersal of seeds. Animal dispersal may sometimes cause an accelerating relation between allocation to female functions and female fertility, because individuals with few fruits would be little visited whereas those with many fruits would disperse many seeds (Bawa 1980; Givnish 1980), or because of wider dispersal of animal-dispersed seeds, generating less competition between sibling seedlings (Lloyd 1980). There is an association between animal dispersed fruits and dioecy, although other possibilities exist for this correlation (Muenchow 1987), and there is no clear evidence for an accelerated gain in fruit removal with increased fruit amounts present on plants (Denslow 1987). Similar arguments have been made for attraction of animal pollinators. If pollinators prefer plants with the largest floral displays, this could lead to an accelerating gain for male allocation (Bawa 1980). Queller (1983) showed that there is indeed a strong advantage in male fertility (as estimated by removal of pollinia) in *Asclepias exaltata*. Better tests of the shape of the male gain curve require estimates of the numbers of seeds actually produced through male function. A recent study of the same species, using this type of data, has found no evidence for an accelerating curve (Broyles & Wyatt 1990).

There may also be sex-specific limits to reproductive success. Some structures (such as a brood pouch for the young of some animals) may have to be produced for any reproductive success to be possible. These costs limit the gain a unisexual form can obtain (equivalent to a diminishing returns gain curve), so that hermaphroditism will be stabilized (Heath 1977). Frank (1987) made the useful distinction between such 'fixed costs' (independent of the number of progeny) and 'packaging costs' that represent the minimum amount of resources that must be invested per progeny. In plants, these packaging costs include pedicels to support the fruits (which it is reasonable to count as female reproductive functions of the maternal plants). These pedicels would also bear the flowers, so that in hermaphroditic species, part of these costs contribute to male function (Lloyd 1989), but in monoecious species one could assign them to male and female functions separately, according to the numbers of flowers of the two sexes and the costs of the two kinds of supporting pedicels. If there are shared reproductive costs that contribute to both sex functions, this should tend to stabilize hermaphroditism. This suggests that allocation to attraction might also stabilize hermaphroditism, and Morgan (1991) has shown, by studying the conditions for stability of a cosexual form to invasion by females, that this is true. This may explain why dioecy is associated with small, inconspicuously coloured flowers (Bawa 1980). Hermaphroditism is also stabilized when the costs attributable to the two sex fertilities come from somewhat different resource pools, because hermaphrodites can then fulfill both functions, i.e. the male–female gain curve should show diminishing returns. One might therefore expect more dioecious species among plants that ripen seeds and fruits over a period of time after flowering has finished.

It is important to remember that since it is the degree of acceleration of the male versus female curve (together with the inbreeding effects) that determines the stability of dioecy to invasion by cosexual forms, an accelerating gain curve for one sex does not guarantee the instability of dioecy, because the other gain curve could be decelerating and make the overall curve unfavourable for the invasion. As discussed above, decelerating curves seem biologically realistic for many plants, at least at the upper part of the allocation range.

Other types of unisexuality may be viewed as allocation biases. Andromonoecy (with individual plants having both hermaphrodite and male flowers) and gynomonoecy (hermaphrodite and female flowers) are biases towards male and female functions, respectively. Among a group of related species (in which the cost per pollen grain, and per unit mass of fruits, should be similar) female bias in allocation, including andromonoecy, should occur in species with the most expensive fruits, and this is indeed found (Bertin 1982; Whalen & Costich 1986). However, the question of why these plants produce female-sterile flowers, rather than simply maturing fruits from few flowers, has not yet been answered. It seems likely that hermaphroditism may be a 'bet-hedging' strategy in the face of unreliable pollination of flowers. The similar problem of gynomonoecy has also not been solved, nor has the difference in the frequncy of these two breeding systems been satisfactorily explained. Gynomonoecy is known from at most 12 families, whereas andromonoecy is widespread. Perhaps the higher allocation of reproductive resources needed for fruiting than flowering more readily permits the evolution of male- than of female-biased allocation at flowering time.

7. CONCLUSIONS

The theory of sex allocation has made a number of interesting predictions, and these have been confirmed in several cases. There is still a need for more empirical data, both for testing the validity of the assumptions that are made in the models, and for finding out whether such ideas as those about gain curves in plants with different ecologies are correct. A particularly valuable type of study would involve estimation of genetic corelations between different reproductive functions. To show trade-offs between different functions, one must show that there is a negative genetic correlation, not merely a phenotypic correlation (Rose & Charlesworth 1981). Without such data, the theoretical basis for the predictions of sex allocation models is weak. Sub-dioecious species would be particularly favourable for such studies, because of the plants with male function in such populations are known to show variability for levels of female function (Delph & Lloyd 1991).

Another important type of data includes studies of the relations between allocation to (or biomass in) different reproductive structures, and the reproductive success generated. There have so far been only a few studies of the gain curve for success through pollen production, in relation either to flower number (Schemske 1980; Schoen & Stewart 1986; Piper & Waite 1988, but see Snow 1989) or attractiveness to pollinators (Bell 1985), and such studies rarely include effects on female fertility that could be used to estimate both male and female gain curves. Part of the reason for the small number of studies is that there is difficulty in estimating male success. Estimates based on numbers of visits by pollinators to flowers, or even on removal of pollen, may not be accurate if there is a nonlinear relation between pollen removed and pollen deposited on the stigmas of other plants, or numbers of seeds sired. Inaccuracy of possibly large magnitude may be common, as shown by Broyles & Wyatt (1990) in a study of *Asclepias exaltata*, in which paternity of seeds could be determined using electrophoretic markers. Even with some inaccuracy, however, such studies are badly needed. Studies of the relation between fruit production and contributions to the progeny in the next generation are also needed. Without these types of information, we cannot assess the plausibility of the ideas that have been proposed for the selective pressures involved in the evolution of dioecy and other forms of unisexuality, and our understanding of breeding system evolution will remain incomplete.

We thank B. Charlesworth for comments on this manuscript. This work was supported by NSF grants BSR 8516629 and BSR 8817976.

REFERENCES

Abrahamson, W. G. 1979 Patterns of resource allocation in wildflower populations of fields and woods. *Am. J. Bot.* **66**, 71–79.

Anderson, G. J. 1979 Dioecious *Solanum* species of hermaphroditic origin as an example of a broad evolutionary convergence. *Nature, Lond.* **282**, 836–838.

Anderson, G. J. & Symon, D. E. 1989 Functional dioecy and andromonoecy in *Solanum*. *Evolution* **43**, 204–219.

Antonovics, J. 1980 Concepts of resource allocation and partitioning in plants. In *Limits to action* (ed. J. E. R. Staddon), pp. 1–35. New York: Academic Press.

Assouad, M. W., Dommée, B., Lumaret, R. & Valdeyron, G. 1978 Reproductive capacities in the sexual forms of the gynodioecious species *Thymus vulgaris* L. *Biol. J. Linn. Soc.* **77**, 29–39.

Atlan, A. A., Gouyon, P.-H., Fournial, T., Pomente, D. & Couvet, D. 1990 Sex allocation in an hermaphroditic plant: *Thymus vulgaris* and the case of gynodioecy. *Evolution.* (In the press.)

Baker, H. G. 1959 Reproductive methods as a factor in speciation in flowering plants. *Cold Spring Harb. Symp. quant. Biol.* **24**, 9–24.

Barrett, S. C. H. & Eckert, C. G. 1990 Variation and evolution of plant mating systems. In *Biological approaches and evolutionary trends in plants* (ed. S. Kawano). New York: Academic Press. (In the press.)

Bawa, K. S. 1980 Evolution of dioecy in flowering plants. *A. Rev. Ecol. Syst.* **11**, 15–39.

Bell, G. 1985 On the function of flowers. *Proc. R. Soc. Lond.* B **224**, 223–265.

Bell, G. & Koufopanou, V. 1986 The cost of reproduction. *Oxf. Surv. Evol. Biol.* **3**, 82–131.

Bertin, R. I. 1982 The evolution and maintenance of andromonoecy. *Evol. Theory* **6**, 25–32.

Broyles, S. B. & Wyatt, R. 1990 Paternity analysis in a natural population of *Asclepias axaltata*: multiple paternity, functional gender, and the 'pollen donation hypothesis'. *Evolution* **44**, 1454–1468.

Brunet, J. 1991 Factors influencing sex allocation among flowers on inflorescences of hermaphroditic plants. (In preparation.)

Burd, M. & Allen, T. F. H. 1988 Sexual allocation strategy in wind-pollinated plants. *Evolution* **42**, 403–407.

Charlesworth, B. 1984 The evolutionary genetics of life histories. In *Evolutionary ecology* (ed. B. Shorrocks), pp. 117–133. Oxford University Press.

Charlesworth, B. 1990 Optimization models, quantitative genetics, and mutation. *Evolution* **44**, 520–538.

Charlesworth, B. & Charlesworth, D. 1978 A model for the evolution of dioecy and gynodioecy. *Am. Nat.* **112**, 975–997.

Charlesworth, D. 1984 Androdioecy and the evolution of dioecy. *Biol. J. Linn. Soc.* **23**, 333–348.

Charlesworth, D. 1989 Allocation to male and female functions in sexually polymorphic populations. *J. theor. Biol.* **139**, 327–342.

Charlesworth, D. & Charlesworth, B. 1978 Population genetics of partial male-sterility and the evolution of monoecy and dioecy. *Heredity, Lond.* **41**, 137–153.

Charlesworth, D. & Charlesworth, B. 1981 Allocation of resources to male and female functions in hermaphrodites. *Biol. J. Linn. soc.* **15**, 57–74.

Charlesworth, D. & Charlesworth, B. 1987 The effect of investment in attractive structures on allocation to male and female functions in plants. *Evolution* **41**, 948–968.

Charnov, E. L. 1979 Simultaneous hermaphroditism and sexual selection. *Proc. natn. Aca. Sci., U.S.A.* **76**, 2480–2484.

Charnov, E. L. 1982 *The theory of sex ratio and sex allocation.* Princeton University Press.

Charnov, E. L. 1988 Hermaphroditic sex allocation with overlapping generations. *Theor. Popul. Biol.* **34**, 38–46.

Charnov, E. L. & Bull, J. J. 1985 Sex allocation in a patchy environment: a marginal value theorem. *J. theor. Biol.* **115**, 619–624.

Charnov, E. L. & Bull, J. J. 1986 Sex allocation, pollinator attraction and fruit dispersal in cosexual plants. *J. theor. Biol.* **118**, 321–326.

Charnov, E. L., Maynard Smith, J. & Bull, J. J. 1976 Why be an hermaphrodite? *Nature, Lond.* **263**, 125–126.

Cohen, D. & Dukas, R. 1990 The optimal number of female flowers and the fruits-to-flowers ratio in plants under pollination and resources limitation. *Am. Nat.* **135**, 218–241.

Cruden, R. W. 1977 Pollen-ovule ratios: a conservative index of breeding systems in flowering plants. *Evolution* **31**, 32–46.

Cruden, R. W. & Lyon, D. L. 1985 Patterns of biomass allocation to male and female functions in plants with different mating systems. *Oecologia* **66**, 299–306.

Cumaraswamy, A. & Bawa, K. S. 1989 Sex allocation and mating system in pigeonpea *Cajanus cajan* (Fabaceae). *Pl. Syst. Evol.* **168**, 59–69.

Darwin, C. 1877 *The different forms of flowers on plants of the same species.* London: John Murray.

Dawson, T. E., King, E. J. & Ehleringer, J. R. 1990 Sex-ratio and reproductive variation in the mistletoe *Phoradendron juniperinum* (Viscaceae). *Am. J. Bot.* **77**, 584–589.

Delph, L. F. & Lloyd, D. G. 1991 Genetic and environmental control of gender in the dimorphic shrub *Hebe subalpina. Evolution.* (Submitted.)

Denslow, J. S. 1987 Fruit removal rate from aggregated and isolated bushes of the red elderberry, *Sambucus pubens. Can. J. Bot.* **65**, 1229–1235.

Eis, S., Garman, E. H. & Ebell, L. F. 1965 Relation between cone production and diameter increment of Douglas fir (*Pseudotsuga menziesii* (MIRB.) Franco), grand fir (*Abies grandis* (Dougl.) Lindl.) and western white pine (*Pinus monticola* Dougl.). *Can. J. Bot.* **43**, 1553–1559.

Frank, S. A. 1987 Individual and population sex allocation patterns. *Theor. Popul. Biol.* **31**, 47–74.

Givnish, T. J. 1980 Ecological constraints on the evolution of breeding systems in seed plants: dioecy and dispersal in gymnosperms. *Evolution* **34**, 959–972.

Godley, E. J. 1964 Breeding systems in New Zealand plants. 3. Sex ratios in some natural populations. *N. Z. Jl Bot.* **2**, 205–212.

Goldman, D. A. & Willson, M. F. 1986 Sex allocation in functionally hermaphroditic plants: a review and critique. *Bot. Rev.* **52**, 157–194.

Grant, M. C. & Mitton, J. B. 1979 Elevational gradients in adult sex ratios and sexual differentiation in vegetative growth of *Populus tremuloides* Michx. *Evolution* **33**, 914–918.

Gregorius, H.-R. 1982 Selection in plant populations of effectively infinite size. II. Protectedness of a biellelic polymorphism. *J. theor. Biol.* **96**, 689–705.

Gregorius, H.-R. & Ross, M. D. 1981 Selection in plant of effectively infinite size: 1. Realized genotypic fitnesses. *Math. Biosci.* **54**, 291–307.

Heath, D. J. 1977 Simultaneous hermaphroditism: cost and benefit. *J. theor. Biol.* **64**, 363–373.

Holsinger, K. E. 1991 Mass action models of plant mating systems: the evolutionary stability of mixed mating systems. (In preparation.)

Horovitz, A. & Harding, J. 1972 Genetics of *Lupinus*. V. Intraspecific variability for reproductive traits in *Lupinus nanus*. *Bot. Gaz.* **133**, 155–165.

Horvitz, C. & Schemske, D. W. 1988 Demographic cost of reproduction in a neotropical herb: an experimental field study. *Ecology* **69**, 1741–1745.

Jing, S. W. & Coley, P. D. 1990 Dioecy and herbivory: the effect of growth rate on plant defense in *Acer negundo. Oikos* **58**, 369–377.

Kakehashi, M. & Harada, Y. 1987 A theory of reproductive allocation based on size-specific demography. *Pl. Species Biol.* **2**, 1–13.

Kawakubo, N. 1990 Dioecism of the genus *Callicarpa* (Verbenaceae) in the Bonin (Ogasawara) islands. *Bot. Mag., Tokyo* **103**, 57–66.

Kawano, S. 1981 Trade-off relationships between some reproductive characteristics in plants with special references to life-history strategy. *Bot. Mag., Tokyo* **94**, 285–294.

Kevan, P. G. 1972 Insect pollination of high arctic flowers. *J. Ecol.* **60**, 831–847.

Kevan, P. G., Eisikowitch, D., Ambrose, J. D. & Kemp, J. R. 1990 Cryptic dioecy and insect pollination in *Rosa setigera* Michx. (Rosaceae), a rare plant of Carolinian Canada. *Biol. J. Linn. Soc.* **40**, 229–243.

Kirkpatrick, M. & Bull, J. J. 1987 Sex ratio selection with migration: does Fisher's result hold? *Evolution* **41**, 218–221.

Liston, A., Rieseberg, L. H. & Elias, T. S. 1990 Functional androdioecy in the flowering plant *Datisca glomerata. Nature, Lond.* **343**, 641–642.

Lloyd, D. G. 1965 Evolution of self-compatibility and racial differentiation in *Leavenworthia* (Cruciferae). *Contrib. Gray Herbarium Harv. Univ.* **195**, 3–134.

Lloyd, D. G. 1975 Theoretical sex ratios of dioecious and gynodioecious angiosperms. *Heredity, Lond.* **32**, 11–34.

Lloyd, D. G. 1977 Genetic and phenotypic models of natural selection. *J. theor. Biol.* **69**, 543–560.
Lloyd, D. G. 1979 Some reproductive factors affecting the selection of self-fertilization in plants. *Am. Nat.* **113**, 67–79.
Lloyd, D. G. 1980 Benefits and handicaps of sexual reproduction. *Evol. Biol.* **13**, 69–111.
Lloyd, D. G. 1982 Selection of combined versus separate sexes in seed plants. *Am. Nat.* **120**, 571–585.
Lloyd, D. G. 1987*a* Parallels between sexual strategies and other allocation strategies. In *The evolution of sex* (ed. S. Stearns), pp. 263–281. Basel: Birkhaüser Verlag.
Lloyd, D. G. 1987*b* Benefits and costs of biparental and uniparental reproduction in plants. *The evolution of sex: an examination of current ideas* (ed. R. E. Michod & B. R. Levin), pp. 233–252. Sunderland, Massachusetts: Sinauer.
Lloyd, D. G. 1987*c* Allocations to pollen, seeds and pollination mechanisms in self-fertilizing plants. *Funct. Ecol.* **1**, 83–89.
Lloyd, D. G. 1989 The reproductive ecology of plants and eusocial animals. In *Towards a more exact ecology* (ed. P. J. Grubb), pp. 185–208. Oxford: Blackwells Scientific Publications.
Lloyd, D. G. & Webb, C. J. 1977 Secondary sex characters in seed plants. *Bot. Rev.* **43**, 177–216.
Lovett-Doust, J. & Cavers, P. B. 1982 Biomass allocation in hermaphrodite flowers. *Can. J. Bot.* **60**, 2530–2534.
Lovett-Doust, J. & Harper, J. L. 1980 The resource costs of gender and maternal support in an andromonoecious umbellifer *Smyrnium olusatrum* L. *New Phytol.* **85**, 251–264.
MacNair, M. R. & Cumbes, Q. J. 1990 The pattern of sexual resource allocation in the yellow monkey flower, *Mimulus guttatus*. *Proc. R. Soc. Lond.* B **242**, 101–107.
Mayer, S. S. 1990 The origin of dioecy in Hawaiian *Wikstroemia* (Thymeleaceae). *Mem. New York Bot. Gard.* **55**, 76–82.
Maynard Smith, J. 1982 *Evolution and the theory of games*. Cambridge University Press.
McGinley, M. A. & Charnov, E. L. 1989 Multiple resources and the optimal balance between size and number of offspring. *Evol. Ecol.* **2**, 77–84.
McKone, M. J. & Webb, C. J. 1988 A difference in pollen size between the male and hermaphrodite flowers of two species of Apiaceae. *Aust. J. Bot.* **36**, 331–337.
Meagher, T. R. 1982 The population biology of *Chamaelirium luteum*, a dioecious member of the lily family: life history studies. *Ecology* **63**, 1690–1700.
Morgan, M. T. 1991 (In preparation.)
Muenchow, G. A. 1987 Is dioecy associated with fleshy fruit? *Am. J. Bot.* **74**, 287–293.
Ornduff, R. 1969 Reproductive biology in relation to systematics. *Taxon* **18**, 121–133.
Ornduff, R. 1987 Sex ratios and coning frequency of the cycad *Zamia pumila* L. (Zamiaceae). *Biotropica* **19**, 361–364.
Pellmyr, O. 1987 Multiple sex expressions in *Cimifuga racemosa*: dichogamy destabilizes hermaphroditism. *Biol. J. Linn. Soc.* **31**, 161–173.
Piper, J. & Charlesworth, B. 1986 The evolution of distyly in *Primula vulgaris*. *Biol. J. Linn. Soc.* **29**, 123–137.
Piper, J. & Waite, S. 1988 The gender role of flowers of broad leaved Helleborine, *Epipactis helleborine* (L.) Crantz (Orchidaceae). *Funct. Ecol.* **2**, 34–40.
Policansky, D. 1981 Sex choice and the size advantage model in jack-in-the-pulpit. *Proc. natn. Acad. Sci. U.S.A.* **78**, 1306–1308.
Preston, R. E. 1986 Pollen-ovule ratios in Cruciferae. *Am. J. Bot.* **73**, 1732–1737.
Queller, D. C. 1983 Sexual selection in a hermaphroditic plant. *Nature, Lond.* **305**, 706–707.
Queller, D. C. 1984 Pollen-ovule ratios and hermaphrodite sexual allocation strategies. *Evolution* **38**, 1148–1151.
Ritland, C. & Ritland, K. 1989 Variation of sex allocation among eight taxa of the *Mimulus guttatus* species complex. *Am. J. Bot.* **76**, 1731–1739.
Robbins, L. & Travis, J. 1986 Examining the relationship between functional gender and gender specialization in hermaphroditic plants. *Am. Nat.* **128**, 409–415.
Rohmeder, E. 1967 Beziehung zwischen Frucht- bzw. Holzerzeugung der Waldebaüme. *Allg. Forstz.* **22**, 33–39.
Rose, M. R. & Charlesworth, B. 1981 Genetics of life-history in Drosophila melanogaster. I. Sib analysis of adult females. *Genetics* **97**, 173–186.
Ross, M. D. & Gregorius, H.-R. 1983 Outcrossing and sex function in hermaphrodites: a resource-allocation model. *Am. Nat.* **121**, 204–222.
Sakai, A. K. & Sharik, T. L. 1988 Clonal growth of male and female bigtooth aspen (*Populus grandidentata*). *Ecology* **69**, 2031–2033.
Sakai, A. K. & Weller, S. G. 1991 Sex expression and sex allocation in subdioecious *Schiedia globosa* (Caryophyllaceae). (In preparation.)
Savage, A. J. P. & Ashton, P. S. 1983 The population structure of the double coconut and some other Seychelles palms. *Biotropica* **15**, 15–25.
Schemske, D. W. 1980 Evolution of floral display in the orchid *Brassavola nodosa*. *Evolution* **34**, 489–493.
Schemske, D. W. & Lande, R. 1985 The evolution of self fertilization and inbreeding depression in plants. II. Empirical observations. *Evolution* **39**, 41–52.
Schlessman, M. A., Lowry, P. P. & Lloyd, D. G. 1990 Functional dioecism in the New Caledonian endemic *Polyscias pancheri* (Araliaceae). *Biotropica* **22**, 133–139.
Schoen, D. J. 1982 Male reproductive effort and breeding system in an hermaphroditic plant. *Oecologia* **53**, 255–257.
Schoen, D. J., Denti, D. & Stewart, S. C. 1986 Strobilus production in a clonal white spruce seed orchard: evidence for unbalanced mating. *Silvae Genet.* **35**, 1109–1120.
Schoen, D. J. & Stewart, S. C. 1986 Variation in male reproductive investment and male reproductive success in white spruce. *Evolution* **40**, 1109–1120.
Schoen, D. J. & Dubuc, M. 1990 The evolution of infloresence size and number; a gamete-packaging strategy in plants. *Am. Nat.* **135**, 841–857.
Silvertown, J. 1987 The evolution of hermaphroditism. An experimental test of the resource model. *Oecologia* **72**, 157–159.
Snow, A. A. 1989 Assessing the gender role of hermaphrodite flowers. *Funct. Ecol.* **3**, 249–250.
Snow, B. & Snow, D. 1988 *Birds and berries*. Calton, Wiltshire: T. and A. D. Poyser.
Spalik, K. 1990 On the evolution of andromonoecy and 'overproduction' of flowers: a resource allocation model. *Biol. J. Linn. Soc.* (In the press.)
Spalik, K. & Woodell, S. R. J. 1991 Regulation of pollen production in *Anthriscus sylvestris* L., an andromonoecious species. (In preparation.)
Stanton, M. L. & Preston, R. E. 1988 Ecological consequences and phenotypic correlates of petal size variation in wild radish, *Raphanus raphanistrum* (Brassicaceae). *Am. J. Bot.* **75**, 526–537.
Stauffer, R. C. 1975 *Charles Darwin's Natural Selection*. Cambridge University Press.
Thomson, J. D. 1989 Deployment of ovules and pollen among flowers within inflorescences. *Evol. Trends Plants* **3**, 65–68.
Thomson, J. D. & Brunet, J. 1990 Hypotheses for the evolution of dioecy in seed plants. *Trends Ecol. Evol.* **5**, 11–16.

Thomson, J. D. & Thomson, B. A. 1989 Dispersal of *Erythronium grandiflorum* pollen by bumblebees: implications for gene flow and reproductive success. *Evolution* **43**, 657–661.

Tuljapurkar, S. 1990 Age structure, environmental fluctuations, and hermaphroditic sex allocation. *Heredity, Lond.* **64**, 1–7.

van Wyk, A. E. & Dedekind, I. 1985 The genus *Eugenia* (Myrtaceae) in southern Africa: morphology and taxonomic value of pollen. *S. Afr. J. Bot.* **51**, 371–378.

Vasek, F. C. & Weng, V. 1988 Breeding systems of *Clarkia* sect. *Phaeostoma* (Onagraceae): I. Pollen-ovule ratios. *Syst. Bot.* **13**, 336–350.

Young, H. J. & Stanton, M. L. 1990 Influences of floral variation on pollen removal and seed production. *Ecology* **71**, 536–547.

Webb, C. J. 1979 Breeding systems and the evolution of dioecy in New Zealand apioid Umbelliferae. *Evolution* **33**, 662–672.

Westergaard, M. 1958 The mechanism of sex determination in dioecious plants. *Adv. Genet.* **9**, 217–281.

Whalen, M. D. & Costich, D. E. 1986 Andromonoecy in *Solanum*. In *Solanaceae: biology and systematics* (ed. W. G. D'Arcy), pp. 284–305. New York: Columbia University Press.

Willson, M. F. 1983 *Plant reproductive ecology*. New York: John Wiley.

The evolution of reproductive strategies: a commentary

J. MAYNARD SMITH

School of Biological Sciences, University of Sussex, Falmer, Brighton BN1 9QG, U.K.

To understand the evolution of any trait, one must know both the selective forces that are operating, and the nature of the genetic variance. Usually we are ignorant of both. One attraction of studying life histories is that the traits under study contribute directly to fitness: if we know the mortality and fecundity schedules of some genotype, we know its fitness. Of course, things are not quite as easy as that, because we need to know mortality and fecundity in the wild, and not in the laboratory, but we are in better shape than we would be if, for example, we were studying courtship or fighting behaviour. Because the contributions of fecundity and longevity to fitness are so direct, the importance of trade-offs is at once obvious. If there were no trade-offs, all organisms would be as fecund as the most fecund, as long-lived as the most long-lived, and would develop as rapidly as the most rapid developer. The problem of trade-offs was mentioned in almost every paper at the symposium, and was a major topic of several – for example, the papers by Charlesworth & Morgan, and by Partridge & Sibly.

There are two ways of approaching the problem of trade-offs. To a geneticist, a trade-off manifests itself as a negative genetic correlation between two traits – for example, long-lived genotypes are less fecund – and such correlations can, at least in principle, be measured. To a physiologist, trade-offs arise from physiological interactions, and can, again in principle, be discovered by experimental manipulation. A geneticist needs to measure the matrix of genetic variances and covariances: given that, and a knowledge of the selective forces, he can predict the course of evolution. The aim of a physiologist is somewhat harder to define. In effect, one wants to know the 'phenotype set', or 'fitness set' in Levins' (1968) terminology: that is, the set of phenotypes that are possible. Given the phenotype set, and a knowledge of the selective forces, one can deduce the optimum phenotype to which the population will evolve. The snag, of course, lies in the word 'possible'. Does this mean 'possible given the genes segregating in the population at the present time', or 'possible given, not only the genes now segregating, but mutations that may arise in the future'? If it means the latter, how long a future? Were the whale and the bat in the phenotype set of the first species of eutherian mammal? In practice, one tends to think, rather imprecisely, of the range of phenotypes that could be produced by selection acting on a species in a few thousands of generations.

At first sight, the case for the genetic approach is a strong one: given the necessary genetic information, we could indeed predict the course of evolution, for a few generations, and assuming a population large enough to make drift unimportant. But the practical difficulties, even for the most favourable organisms, are formidible. We need to measure phenotypes in the wild. We must measure genetic correlations: if some individuals have a more favourable environment than others, we may find spurious positive correlations between fitness traits. We need to measure the additive component of the genetic variance: if members of a population vary in degree of homozygosity, this could give a misleading genetic correlation between sibs. Perhaps most serious of all, there is no reason to expect genetic correlations to be constant within a species; for example, correlations can arise from linkage disequilibrium caused by recent selection, and be very transitory. Indeed, one can argue that the only correlations likely to be sufficiently stable to be interesting arise from physiological interactions that could more easily be discovered by direct experimental manipulation.

There is one case in which genetic and physiological methods have been applied to the same problem, happily with consistent results. Rose & Charlesworth (1980) showed, in *Drosophila melanogaster*, a negative genetic correlation between adult female longevity, and fecundity when young. This fits with the results of physiological manipulation (Lamb 1964; Maynard Smith 1958), admittedly on a different species, *D. subobscura*, showing that a range of treatments that reduce the rate of egg-laying (including raising ovariless females from mothers homozygous for the gene *grandchildless*, and giving young adults 5000 rads of X-rays) prolong adult life. The moral, I think, is that genetic and physiological approaches can be complementary. An attempt to measure the genetic covariance matrix in the wild does not seem to me to be a sensible enterprise, but the use of genetic analysis to reveal pleiotropisms, which must have a physiological basis, can be valuable.

I first met the idea of trade-offs when working in aircraft design. If, for example, you want an aeroplane to fly fast, you give it small wings, and pay for the high speed with a long take-off run. This is all predictable, because there is a theory that relates lift and drag, and hence take-off speed and top speed, to wing area. One reason for our difficulties in life-history theory is that we do not have a comparable quantitative theory either of senescence or of development rate. The absence of a theory of development rate can be illustrated by our inability to answer the following questions. What places an upper limit on the rate of growth? Why is it that bacteria, in optimal conditions, can double in 20 minutes, whereas a mammalian cell, at best, takes 8 hours? It is possible to suggest answers to these questions (Maynard Smith 1969), but they are no more than plausible guesses. We are in similar difficulties over senescence. For example, suppose an animal has been selected to have a long adult life before senescence sets in; is it necessary (for example, because of the need for accurate molecular replication) that it should have a long development time? Again, we can speculate, but we do not know. The fact that, across taxa, there is a positive correlation between longevity and development time does not by itself tell us why the correlation exists.

This brings me to my final topic, the comparative method, discussed at this symposium by Harvey and Keymer. In recent years the method has been transformed: it is no longer possible to dream up a theory, and then to search the literature for a few species that fit the predictions. Through the introduction of phylogenetic reasoning and statistical analysis, we now have powerful tools for testing theories about evolution. I am not sure, however, that we are yet very good at using them. Suppose that we are satisfied that traits A and B are correlated across taxa, in the sense that, when one arises in an evolutionary lineage, the other is likely to do so. We are still a long way from a causal explanation. The two traits may be physiologically independent adaptations to the same ecological feature (herbivorous mammals are often both hypsodont and have a long alimentary canal); the two traits may be independent adaptations to two features of the environment that commonly co-occur (cold places are often covered by snow, so that mammals with thick fur are often white); trait A may be an ecological adaptation, and B a secondary adaptation entailed by A (large land mammals have relatively stout legs); two traits may be demographically linked (animals with a high juvenile mortality must have high fecundity); there may be chemical or physical constraints that cause A and B to occur together (because of the law of levers, animals that dig tend to run slowly). Doubtless there are other possibilities. The snag is that, when thinking about life histories, we lack the kind of theory (except, of course, for demography) that would enable us to make predictions like 'large land animals will have relatively stout legs'. One of the most encouraging features of the symposium, shown, for example, in the relation between the papers by Charnov and by Harvey & Keymer, was the indication that people are seeking a tie-up between models on the one hand and comparative data on the other.

REFERENCES

Lamb, M. J. 1964 The effects of radiation on the longevity of female *Drosophila subobscura*. *J. Insect Physiol.* **10**, 487–497.

Levins, R. 1968 *Evolution in changing environments*. Princeton University Press.

Maynard Smith, J. 1958 The effects of temperature and egg-laying on the longevity of *Drosophila subobscura*. *J. exp. Biol.* **35**, 832–842.

Maynard Smith, J. 1969 Limitations on growth rate. *Symp. Soc. gen. Microbiol.* **19**, 1–13.

Rose, M. & Charlesworth, B. 1980 A test of evolutionary theories of senescence. *Nature, Lond.* **287**, 141–142.